Global Magic

Palgrave Studies in Anthropology of Sustainability

Our series aims to bring together research on the social, behavioral, and cultural dimensions of sustainability: on local and global understandings of the concept and on lived practices around the world. It will publish studies focusing on ways of living, acting and thinking which claim to favor the local and global ecological systems of which we are part, and on which we depend for survival. Political pressure surrounding sustainable resource governance shapes regimes of measurement and control and the devolution of risk and responsibility. Scientific cultures of sustainability are generated out of concern over the need for "green" technologies and materials. Popular discourses of scarcity of resources or capital increasingly lead to challenges to cosmopolitan and egalitarian ideals (human rights, the welfare state), fed by fears over the sustainability of social systems and civilizations in the face of global change. Meanwhile an array of social and cultural transformations is occurring that seek to offer ways to live (and produce, consume . . .) more sustainably. Calculations of sustainability raise questions of value—a vexed political affair. An anthropological approach will help understand these emerging phenomena.

Series Editors: Marc Brightman and Jerome Lewis

Titles:

Global Magic: Technologies of Appropriation from Ancient Rome to Wall Street by Alf Hornborg

Global Magic

Technologies of Appropriation from Ancient Rome to Wall Street

Alf Hornborg

GLOBAL MAGIC
Copyright © Alf Hornborg 2016

All rights reserved. No reproduction, copy or transmission of this publication may be made without written permission. No portion of this publication may be reproduced, copied or transmitted save with written permission. In accordance with the provisions of the Copyright, Designs and Patents Act 1988, or under the terms of any licence permitting limited copying issued by the Copyright Licensing Agency, Saffron House, 6-10 Kirby Street, London EC1N 8TS.

Any person who does any unauthorized act in relation to this publication may be liable to criminal prosecution and civil claims for damages.

First published 2016 by
PALGRAVE MACMILLAN

The author has asserted his right to be identified as the author of this work in accordance with the Copyright, Designs and Patents Act 1988.

Palgrave Macmillan in the UK is an imprint of Macmillan Publishers Limited, registered in England, company number 785998, of Houndmills, Basingstoke, Hampshire, RG21 6XS.

Palgrave Macmillan in the US is a division of Nature America, Inc., One New York Plaza, Suite 4500, New York, NY 10004-1562.

Palgrave Macmillan is the global academic imprint of the above companies and has companies and representatives throughout the world.

Hardback ISBN: 978-1-137-56786-4
E-PUB ISBN: 978-1-137-56788-8
E-PDF ISBN: 978-1-137-56787-1
DOI: 10.1057/9781137567871

Distribution in the UK, Europe and the rest of the world is by Palgrave Macmillan®, a division of Macmillan Publishers Limited, registered in England, company number 785998, of Houndmills, Basingstoke, Hampshire RG21 6XS.

Library of Congress Cataloging-in-Publication Data

Names: Hornborg, Alf.
Title: Global magic : technologies of appropriation from ancient Rome to Wall Street / Alf Hornborg.
Description: New York, NY : Palgrave Macmillan, [2016] | Includes bibliographical references and index.
Identifiers: LCCN 2015034895 | ISBN 9781137567864 (hardcover : alk. paper)
Subjects: LCSH: Economic anthropology. | Technology—Anthropological aspects. | Technology and civilization.
Classification: LCC GN448 .H66 2016 | DDC 306.3—dc23 LC record available at http://lccn.loc.gov/2015034895

A catalogue record for the book is available from the British Library.

To Anne-Christine

Contents

Acknowledgments — ix

Introduction — 1

1 The Ecology of Things: Artifacts as Embodied Relations — 9

2 Land, Energy, and Value in the Technocene — 17

3 The Magic of Money — 37

4 Empires, World-Systems, and Expanding Markets — 53

5 Money as Fictive Energy: Unraveling the Relation between Economics and Physics — 75

6 Agency, Ontology, and Global Magic — 93

7 The Political Ecology of Technological Utopianism — 113

8 Redesigning Money to Curb Globalization and Increase Resilience — 129

Conclusions: Money, Technology, and Magic — 151

Notes — 165

References — 175

Index — 193

Acknowledgments

For permission to reproduce revised versions of texts previously published elsewhere, I thank the following journals and publishers:

Chapter 2 integrates parts of chapters published in the volumes *Cultures of Energy: Power, Practices, Technologies*, edited by Sarah Strauss, Stephanie Rupp, and Tom Love, pp. 41–59 (Left Coast Press, 2013) and *The Anthropocene and the Global Environmental Crisis*, edited by Clive Hamilton, Christophe Bonneuil, and François Gemenne, pp. 57–69 (Routledge, 2015).

Chapter 5 includes parts of an article published in the Elsevier journal *Ecological Economics*, volume 105 (2014), pp. 11–18.

Chapter 6 is a revised version of an article published in *HAU: Journal of Ethnographic Theory*, volume 5, issue 1 (Spring 2015), pp. 47–69.

Chapter 7 includes parts of a chapter published in the volume *Green Utopianism*, edited by Johan Hedrén and Karin Bradley, pp. 76–97 (Routledge, 2014).

Chapter 8 includes parts of an article published in the journal *Resilience: International Policies, Practices and Discourses*, volume 1, issue 2 (2013), pp. 116–129.

Innumerable people have helped shape the ideas I present in this book. Naming individual colleagues would yield cumbersome lists numbering in the hundreds, so I limit myself to expressing my gratitude in general. For companionship and stimulating conversations over many years I am grateful to friends in the Human Ecology Division and the Department of Human Geography, Lund University, and to co-participants in the recent research projects LUCID, EJOLT, FESSUD, Time, memory and representation, and Beyond GDP growth. Many thanks to the Swedish Research Council FORMAS, the Bank of Sweden Tercentenary Foundation, and the EU Seventh Framework Program for funding these projects, which has made it possible for me to write this book.

Acknowledgments

I also want to take the opportunity to express my appreciation to all the fellow anthropologists and many friends from other disciplines who I have encountered at conferences and seminars in various parts of the world, whose perspectives and advice have helped me gain more precision and clarity. This also applies to all those students over the years whose critical questions have prompted me to elaborate and sharpen my arguments. I sense a strong moral support from the wide community of academics and activists whose deepest concern is to grasp the logic driving increasing global polarization and environmental degradation, and to remind ourselves of its devastating consequences for all the victims—human and nonhuman—of global magic.

On a more personal level, my deepest gratitude goes to the little circle of people who continue to be my main source of trust and meaning in life: my wife, children, and grandchildren.

Yxnevik in September 2015
Alf Hornborg

Introduction

In 1914, on the island of Saibai in the Torres Strait of southern Melanesia, local people prophesied that a steamship would soon arrive from beyond the horizon bringing the spirits of dead ancestors and great quantities of desirable cargo to the natives and thus transforming the increasingly unequal relation between European colonizers and colonized Melanesians. For several decades, such "cargo cults" stirred throughout Melanesia. Similar ideas can be traced back to the 1880s, but they became particularly prominent in the decade following the Second World War, after which the anticipated ships were often replaced by airplanes. The coveted goods, including everything from tinned food to flashlights, rifles, refrigerators, and automobiles, were held to be manufactured by dead ancestors. To prepare for their arrival, Melanesians constructed superficial copies of docks, airstrips, warehouses, and radio masts using whatever materials they could muster.

Such local interpretations of the material affluence of European colonizers are not difficult to understand. As Peter Worsley (1970 [1957]: 107) explains,

> As far as the natives were concerned, the Whites received the goods by steamer from unknown parts; they did not manufacture them, and merely sent pieces of paper back. They did no apparent work themselves, yet refused to share their fortune, forcing the natives to work long and hard for a return of a small proportion of the goods they themselves obtained with such ease and in such profusion. Who made these goods, how and where, were mysteries—it could hardly be the idle White men. It was the natives who did all the manual work. If the goods were made in some unknown land, they must, then, be made by the spirits of the dead.

Worsley concludes that the cargo cults are not to be seen "as an irrational flight from reality or a regression from the present into the past but as a quite logical interpretation and criticism of a European-controlled order that itself is full of contradictions which seem inexplicable in rational terms to the natives" (250). Although the beliefs and practices of these nonmodern Melanesians have generally struck modern Europeans as magical, pathetic, and ridiculous, Worsley underscores that "the failure of magical action is . . . a function of limitations of knowledge which are socially conditioned, not a failure to use rational procedures" (277).

Contemporary anthropology tends to agree that it would be inaccurate to characterize the adherents of cargo cults as "irrational," but there seems to be a consensus that their behavior was the result of an inadequate familiarity with the operation of the modern world economy. For instance, Marvin Harris (1971: 567) asserts that "the confusion of the Melanesian revitalization prophets is a confusion about the workings of sociocultural systems. They do not understand how the productive and distributive functions of modern industrial society are organized." The cosmology underlying the cargo cults, in this view, was a local perspective constrained by the limited horizons of indigenous people insufficiently incorporated in the global economy—the world viewed from premodern Melanesia. The implicit corollary is that a correct understanding of the operation of modern industrial society is the prerogative of modern people inhabiting the "developed" countries at the core of the world-system.

This book challenges such assumptions. It argues that the worldview established in nineteenth-century Europe is as constrained by cultural categories and limited horizons as that of premodern Melanesia. Although there can be no question that Europeans have been in a vastly better position to strategically utilize and control industrialism and the world economy than the indigenous peoples whom they have conquered on other continents, this is not equivalent to saying that the predominant European understanding of the operation of the industrial world order is complete or accurate. To be the promoters and beneficiaries of industrialization is not necessarily to be aware of its global prerequisites. The categories and models of mainstream economics are as cultural as the premodern worldviews which they have displaced (Gudeman 1986)—they represent the world viewed from nineteenth-century Europe. The provisional efficacy of a given worldview—whether geared to slavery, the pursuit of bullion, or the combustion of fossil fuels—is not tantamount to its verification as a robust representation of the conditions of economic expansion.

For many anthropologists, such insights are the ultimate raison d'être of their discipline. To turn the anthropological gaze back at the society from which it came, identifying its cultural assumptions, idiosyncrasies, and blind spots, is a potent form of political critique (Marcus and Fischer 1986). Unless we subscribe to some version of full-fledged cultural relativism, it makes it possible for us to reveal materially significant but culturally invisible aspects of the social systems of which we are a part. The readiness to accept that our own established modes of thought may repress or mystify circumstances that impinge on our lives is an inescapable implication of anthropology. To expose such culturally invisible conditions of our existence is much facilitated by juxtaposing our own conceptions with those recorded in very different cultural contexts. Of particular interest in this book is the great cross-cultural variation in how economic and human–environmental relations are conceived.

From the Mexican village of Tzintzuntzan, anthropologist George Foster (1965) reported a belief, widely held among people in peasant societies throughout the world, positing that any affluence enjoyed by one person inevitably comes at the expense of someone else. This zero-sum view of the world was labeled "the image of limited good" and regarded by most modernists at the time as a cultural misconception standing in the way of development. Today, increasing economic polarization, resource exhaustion, and climate change appear to be vindicating the intuitions of those peasant populations, but now transposed from the village to the global scale. In accordance with the anthropological approach sketched above, we should now be as ready to scrutinize and query the mainstream "image of *un*limited good" (Hornborg 1992; Trawick and Hornborg 2015) as a remarkable and misguided cultural feature.

It should be evident, however, that such a "symmetric anthropology" (Latour 1993) must remain very far from symmetric in political terms. To depict nonmodern Melanesians or Mexican villagers as culturally confused is a very different project from subjecting mainstream economics to the same treatment. Applying the tools of cultural analysis to established Western cosmology is to challenge the conceptions which reproduce contemporary power structures. It would be naïve to suggest that cultural analysis alone could subvert those structures. However, as the world order that baffled the Melanesians and today vindicates the Mexican villagers seems in line for crises of several kinds, we may soon find ourselves in need of revised understandings of the conditions of that world order. Over the next few decades, rising concerns with sustainability, energy, climate, and financial solvency may provide a crucial role for cultural analysis in delivering adequate new understandings of the world order that solidified in the nineteenth century. Before too long, such new understandings may be in high demand among politicians and ordinary citizens alike.

We have good reasons to scrutinize mainstream Western cosmology as a cultural system. Cosmologies tend to rationalize the shortcomings of the social order. Among the most obvious shortcomings of the current world order is its inclination to generate abysmal inequalities and ecologically disastrous patterns of consumption and resource use, and yet our mainstream discourse tends to represent these conditions merely as the deplorable but unavoidable side effects of progress. As we look back at the systems of slavery and colonialism that propelled European expansion in the nineteenth century, it is evident to us that racism at the time was a cultural and ideological rationalization of the exploitation of non-European peoples. At the level of conventional public discourse, at least, it is no longer politically correct to regard non-European peoples as intrinsically inferior. Yet exploitation and global inequalities have continued in new forms and increased tremendously since the abolishment of slavery and the liberation of former colonies. Apparently there is something about our current world order that not only continues to generate rising inequalities but also rationalizes them as normal, expectable, and natural consequences of the operation of the world economy. But ideologies that buttress power structures are able to serve such functions precisely by presenting themselves as unquestionable knowledge. Only retrospectively do these functions protrude as evident. If hindsight tells us that racism was an ideology that rationalized slavery and colonialism in the nineteenth century, it seems difficult to accept that the aggravated inequalities of today's world are objectively accounted for by the economic cosmology of our time. In the same way that racism can today be exposed as the cultural prop for slavery and colonialism, we have good reasons to critically scrutinize the cultural assumptions of mainstream economics as rationalizations of global inequalities and ecological degradation.

To people persuaded by our conventional worldview that human history, by and large, is a story of progress, it may seem unwarranted to expect that same worldview to mystify or justify environmental destruction and human impoverishment. At first glance, it does seem unreasonable to deny that the quality of life of most humans has improved over the past few millennia, at least in material terms. It is thus not difficult to view the forces that propel the development of global human society as fundamentally benevolent and to question the urge to expose the occurrence of ulterior motives, hidden agendas, and denied adversities. But quality of life, including environmental quality, is very unevenly distributed among the seven billion people on Earth. The most affluent populations of the world, who can often trace their affluence historically to European expansion in the nineteenth century, are generally able to keep the adverse aspects of world society outside their immediate field of vision. Nevertheless, poverty, malnutrition, illness, violence, repression, and environmental degradation in other sectors of society are as much the adverse side of their modern affluence as their own diffuse feelings of alienation and disorientation. The extent to which material progress is a local experience, contingent on the zero-sum logic of more extensive social systems, is a matter that can be investigated through transdisciplinary research combining social-science understandings of power, exchange, and ideology with natural-science methods for tracing asymmetric resource flows and the uneven distribution of ecological degradation. The very feasibility of displacing work and environmental loads to other populations is a consequence of the human use of symbols and artifacts, as discussed in chapters 1 and 2. Such perspectives on progress can be applied to any supralocal system of exchange, from the aggrandizement of ancient emperors to capital accumulation among merchants, industrialists, and financial speculators. This book seeks to show that what these systems of exchange have in common is precisely the urge to displace work and environmental loads to other populations. In this sense, they are all modes of *appropriation*. They are all founded on the appropriation of human labor and the products of natural space elsewhere.

This conclusion naturally prompts us to reassess the conventional notion of progress. But rather than attempt to detail what a more egalitarian and sustainable notion of progress might entail, the primary objective in this book is to dissolve the illusory boundary between culture and science. The European narrative of the Enlightenment has served to distinguish between nonmodern cosmologies constrained by false assumptions and thus amenable to cultural analysis, on the one hand, and modern accounts of the world systematically pursuing the truth, on the other. Much as Karl Marx understood the operation of ideology, however, anthropology is able to show that modern disciplines such as economics appear to be systematically *obscuring* the truth. This is not to attribute malicious, conspiratorial intentions to economists but merely to note how discourses tend to exclude or suppress perspectives that would undermine the professional efficacy and self-esteem of specific categories of practitioners.

Moreover, as we shall explore particularly in chapter 5, even the most critical alternatives to mainstream understandings of industrial society, such as Marxism, risk being constrained by concepts and implicit assumptions shared with conventional approaches. A particularly important source of confusion in these discussions has been the relation between material parameters such as energy, on the one hand, and notions of economic value, on the other. Underlying much of the classical Marxist theories of surplus value and declining rates of profit, I shall argue, is a compelling but largely intuitive concern with embodied energy and diminishing returns.

A significant aspect of conducting a cultural analysis of modern industrial capitalism is to abandon assumptions about a dichotomy between "our" rationality and "their" magic. As we saw regarding the cargo cults, magic can be rational and vice versa. The failure of cargo magic was a consequence of limited knowledge about the conditions which made a certain social order possible. But lack of sufficient knowledge is a recurrent state of affairs in human history and ubiquitous in societies facing collapse. It is thus essential to begin by delineating a definition of "magic" that makes the concept more useful than simply a category for condescendingly dismissing forms of rationality that, to modern people, seem uninformed. Magic is not merely a practice constrained by the absence of objectively efficacious knowledge but a particular kind of social strategy for achieving specific ends. As defined here, magic hinges on the attribution to certain objects of an agency that is actually contingent on human perceptions rather than on the physical properties of the objects themselves, but that to humans *appears* to be independent of their perceptions. This understanding of magic accommodates not only our ordinary image of the magician's art but also the sense in which Marx revealed the role of money in modern society by characterizing it as an example of "fetishism." When Michael Taussig (1980) reports how nonmodern people in Colombia resort to magical rites such as baptizing money in an effort to increase their income, he illustrates the irony of applying an inadequate kind of magic to an artifact which is itself magical, but the secret control of which is beyond their reach. Throughout this book, and most explicitly in chapter 6, the point I ultimately want to make is that the globalized technologies that began to organize world society in the late eighteenth century can be reconceptualized as a form of magic.

The history of the anthropological notion of magic has been traced elsewhere (for instance, Tambiah 1990) and shall not detain us here. This notion has often served as a contrast to science, illustrating the European distinction between premodern superstition and the modern pursuit of truth. For some anthropologists, notably Bronislaw Malinowski (1954), it signifies a mode of thought and practice that all people are prone to adopt under particular psychological circumstances. Such considerations, however, are not addressed in this book. Here the notion of magic is used in contrast to our conventional concept of technology, as one of two diametrically opposite ways of using artifacts. In both cases, artifacts are believed to have agency—that is, to be able to act so as to achieve a purpose of some kind. The difference between magic and technology has been obvious to most Europeans since the eighteenth century—whereas magic falsely attributes agency to objects on the basis of misguided assumptions, technology accurately acknowledges the capacities of objects to achieve given purposes based on their inherent physical properties. The distinction was a central aspect of the Enlightenment and the Industrial Revolution. From now on, the agency of objects was understood to be contingent only on the design of their physical constitution, rather than on the perceptions or conceptions of humans. Nineteenth-century Europeans frequently ridiculed nonmodern peoples in the colonies for mistaking their superior technologies for magic, that is, for not understanding the difference.

Although not applied to technology, Marx's concept of fetishism illuminates a habit of thought that became entrenched through the Enlightenment and Industrial Revolution. We may refer to it as the abandonment of relationism. As explained in chapter 1, the concept of relationism here denotes the acknowledgment that seemingly bounded material objects should be understood as the products of wider and intangible fields of relations. Among nonmodern, indigenous peoples throughout the world, it is generally recognized that a human or nonhuman organism is a manifestation of the webs of semiotic and material flows that constitute societies and ecosystems. Eighteenth- and nineteenth-century Europeans, however, became obsessed with the *internal* constitution of objects such as organisms and machines. To trace the anatomy of the organism and the blueprint of the machine was regarded as a sufficient account of their operation, to the exclusion of the external flows that are as incontrovertibly necessary for their existence. The Enlightenment illuminated the internal constitution of living and nonliving things, but obscured the significance of their external relations. In the most general sense, this explains how one of the most pervasive features of modernity is the alienation of the human individual from the environment. But it also explains why modern technology is perceived as independent from the global resource flows that sustain it, which ultimately means that it is perceived as independent of the world economy. The science of ecology and the environmental movement have struggled to resurrect the insight that humans cannot be understood as separate from their environment, and similar observations have been made regarding the resource requirements and ecological impacts of hazardous technologies. However, the illusion that technological progress is propelled primarily by ingenuity, independent of prevailing exchange rates on the world market, tends to persist. This illusion, which can be referred to as technofetishism, disregards the extent to which the agency of technological objects is ultimately contingent on the perceptions and strategies of humanmarket actors. In other words, it disregards how, at the global level, the distinction between technology and magic dissolves. Locally, it may seem perfectly adequate to account for a machine by referring to its design, but from a global perspective, such an account is as insufficient as it would be to explain what keeps an organism alive by referring only to its anatomy.

I am well aware that the topics and perspectives dealt with in this book may seem diverse and disparate, ranging from economic anthropology, archaeology, history, and ethnography to thermodynamics, systems theory, financialization, Marxism, Actor-Network Theory (ANT), magic, semiotics, resilience theory, photovoltaic energy, and complementary currencies. Instead of apologizing I underscore the importance of developing transdisciplinary perspectives on the global human predicament. For decades I have deplored how, in pursuing conventional intradisciplinary careers, researchers run a serious risk of succumbing to disciplinary myopia. Doctoral students are rapidly "disciplined" into applying specialized discourses, terminologies, and methodologies that constrain their ability to retain the holistic perspectives with which they may have begun their studies. The sheer quantity of specific concerns that have been explored *within* each discipline is enough to discourage transdisciplinary excursions. To try to develop the kind of integrated worldviews that we can detect among classical and Renaissance philosophers would today risk either being dismissed as superficiality or leading to information overload. The result is that academic knowledge production selects for very specialized concerns, largely disconnected from overarching questions about the prospects of humankind which I believe people in general want to see answered. There are tens of thousands of researchers worldwide who consider themselves committed to understanding the challenges of sustainability, yet few of them are prompted to develop an understanding of the general problems facing an expanding and polarizing global economy confronted with a finite biosphere. Like the blind men exploring different parts of the elephant, none of them is in touch with more than a very limited aspect of the total phenomenon generating the category of data they are equipped to register.

This fragmentation of knowledge production should not be interpreted as a consequence of some kind of intentional conspiracy, but simply as the inexorable result of intradisciplinary selection processes encouraging specialization and the narrowing of questions asked. Yet the unfortunate implication of this logic is that unsustainable and inequitable structures and practices are largely left intact, illustrating the kind of subtle relations between knowledge and power that were identified by Michel Foucault. There are literally unlimited quantities of topics to which a doctoral thesis can be devoted, and all too often even senior researchers tend to continue to stick to the narrow questions to which they were advised to confine their thesis. The proliferation of specialists thus increases exponentially, and genuinely novel perspectives risk drowning in a flood of publications which no one has time to survey. The only possible way of countering such *dis*integration of knowledge is to consciously promote and engage in transdisciplinary research which aspires to integrate perspectives from different disciplines and reassemble overarching concerns with sustainability and justice. I believe that anthropology can serve as an excellent point of departure for such integration, but we must exert ourselves to extend the relevance of our concerns far beyond the boundaries of conventional anthropology. The diversity of disciplinary discourses suggests a confusion of tongues reminiscent of ancient Babel, but as I argue particularly in chapter 6, rather than succumb entirely to the linguistic turn we must keep in mind that all these voices refer to a single and common world.

CHAPTER 1

The Ecology of Things: Artifacts as Embodied Relations

In the fifteenth century BC, queen Hatshepsut of Egypt had two huge granite obelisks carved in honor of her divine father, which were transported from Aswan to Karnak. Stone reliefs at Hatshepsut's mortuary temple Deir el-Bahri show the obelisks being conveyed by ships along the Nile. One of the obelisks stands 30 meters high and is estimated to weigh around 320 tons. The reliefs are strikingly similar to modern blueprints. They represent in informative detail the ancient technology of moving obelisks, complete with pulleys, ropes, and great numbers of rowers. The 3,500-year-old images can help us distinguish analytically between engineering and energy sources. It is evident that the technology of monumental architecture three-and-a-half millennia ago required specialized technical knowledge. Although adapted to the practicalities of harnessing slave labor, ancient Egyptian engineering is as analytically distinguishable from slavery as modern engineering knowledge is distinguishable from economic access to fossil fuels. "Technology" in the sense of expert knowledge is as much a necessary condition for transporting ancient obelisks as it is for modern air travel, but in neither case is it a sufficient condition. Without fossil fuels, our technological knowledge would be as powerless as queen Hatshepsut would have been without slaves.

Technologies, in other words, have two aspects. One is the ingenuity underlying technical design and generally celebrated as the primary source of technological progress. The other is the societal arrangement through which that design can be applied so as to harness a particular source of energy. The two aspects constitute and reinforce each other. Just as technical knowledge defines what can be utilized as an energy source, energy sources define what can serve as technical knowledge. But energy sources are not just out there, waiting to be exploited. In order for slaves or fossil fuels to serve as an energy source for someone, they have to be made available for him or her to exploit. The societal arrangements by which energy sources are made available to different individuals or groups are what we conventionally refer to as the economy. Economies can be defined as modes of distributing resources and risks in human populations. They are universally legitimized by cosmological systems justifying particular patterns of distribution by reference to moral principles. In this abstract sense, the societal function of modern economics is equivalent to the ideology accompanying ancient Egyptian slavery. If a reader should find the comparison objectionable, we might respond by observing that the global inequalities organized by modern economics are considerably more severe than those of ancient Egypt. But the main point to be made here is that "economies" are generally excluded from the definition of "technologies," even though the former are crucial conditions for the existence of the latter.

If we consider other animal species, we can nowhere find intraspecific inequalities even remotely similar to those generated within human societies. This unique inclination of human populations toward complex structures of inequality is closely connected to another uniquely human feature: the anchoring of social relations to extrasomatic points of reference such as language, symbols, and artifacts. Collaborating with primatologist Shirley Strum in a study of baboon behavior, Bruno Latour noted long ago that this is the fundamental difference between the social life of baboons and that of humans (Strum and Latour 1987). Latour went on to theorize the role of artifacts in organizing human social relations, asserting that the things we engage with tend to shape our relations and our modes of thinking about the world. His so-called Actor-Network Theory recognizes artifacts as "actants" that possess autonomous agency just as humans do. Certainly, language, symbols, and artifacts help to organize and buttress social structures, but they will be treated here as props employed in the service of human intentions and strategies, rather than as autonomous agents. A significant perspective contributed by Latour, however, is the understanding of technologies as systems of artifacts that contribute to the organization of human social relations. As technologies are always embedded in economies, Latour's perspective should apply no less to systems of exchange. Beyond the organizing power of language, monetary tokens, gifts, commodities, and technologies are the very stuff of human society.

Such artifacts can be perceived in very different terms, however. While premodern valuables and gifts were understood to embody lasting social relations, modern money and commodities tend to be perceived as autonomous objects severed from the exchange relations that they reflect. In economic anthropology, the contrast is often mentioned between Marcel Mauss's reflections on the fact that premodern Maori experienced gifts as animated by the person of the giver and Marx's observation that modern Europeans perceive commodities as dissociated from their anonymous manufacturers. In modern society, Marx noted, relations between people are represented as relations between things. Marx referred to this exaltation of things, stripped of their social context, as fetishism. He coined the concepts of "money fetishism" and "commodity fetishism" to denote the tendency in modern society to exchange artifacts perceived as autonomous agents, because they are disembedded from social relations.

It will be observed that Marx's and Latour's approaches to artifactual agency are at odds with each other. From Marx's perspective, Latour's ANT is tantamount to an endorsement of fetishism, whereas from Latour's perspective Marx's critique of fetishism is an unwarranted denunciation of practices (the animation of nonliving objects) which are ubiquitous among humans (Latour 2010). As indicated above, the argument in this book is aligned with Marx's position that the autonomous agency of artifacts is an illusion. Regardless of how they are perceived, artifacts are objectively generated in systems of relations which can be investigated by social and natural sciences. Latour is obviously right in demonstrating that social processes are results of the specific ways in which humans and artifacts interact, but we also have to consider that artifacts in themselves are generated in social processes involving exchanges not only of information but also of matter and energy. The conceptual detachment of objects from the relations which spawned them is a peculiarity of post-Enlightenment modernity. It implies an abandonment of a widespread premodern ontology that we may call relationism. Few anthropologists have challenged us to rethink ecology in relationist terms, but two prominent anthropological champions of relationism deserve mention here.

Relationist Approaches in Anthropology

The first anthropologist to mention in this context is Gregory Bateson (1972). Although his research topics ranged widely from ethnography, psychiatry, art, and epistemology to evolution, ecology, and animal behavior, he applied a remarkably consistent theoretical framework to his various fields of study. An attempt to summarize this framework would emphasize its recognition of the extent to which concrete forms and patterns in culture and biology are the products of wider fields of interaction. Whether details of indigenous New Guinea ritual, schizophrenia, alcoholism, military strategies, evolutionary adaptation, or the playful wrestling of young mammals, the phenomena Bateson focused on were consistently interpreted as the outcome of communicative processes within less tangible fields of relations. This approach is entirely consonant with his pioneering development of the perspectives of cybernetics, or systems theory. Cybernetics originated as an attempt to incorporate features of biological systems in the design of technologies (cf. von Bertalanffy 1968). Bateson was aware that the form of each individual component of a living system develops in relation to the webs of interaction in which it is embedded. His wide-ranging studies in anthropology, psychiatry, and biology provided a corrective to the conventional inclination to explain behavioral and physical forms with reference to their internal drives and constitutions, rather than to their external relations. Moreover, he realized early on that the Western preoccupation with the inner operation of bounded forms was an epistemological fallacy. Its most problematic consequence was the notion of humans as entities apart from their environment. Although, by now, the science of ecology and the environmental movement have helped raise a general awareness of how dependent humans are on their natural environment, it remains difficult for most modern people to conceive of organisms as vortices of matter, energy, and information reproduced by socioecological flows. Even more difficult to digest, the current argument would add, is that this also applies to artifacts. The conditions by which biomass is maintained also apply to what can be referred to as technomass.

The second anthropologist who should be mentioned here is Tim Ingold (2000). Similarly wide ranging in interests, his general approach is highly influenced by Bateson's. Also straddling biology and anthropology, Ingold applies what he calls "relational thinking" to his accounts of phenomena as diverse as perception, movement, animal behavior, evolutionary processes, learning a skill, craftsmanship, and technology. He has shown how the constitution and behavior of living things generally emerge as aspects of the relations in which they are embedded, rather than being determined by some preexistent program. Like Bateson, he notes that mainstream Western science has encouraged representations of the world in which human and nonhuman organisms, and their products, are objectified and disembedded from their relational contexts. Ingold specifically demonstrates how the emergence of modern technology can be seen as an externalization of work from human organisms onto machines. He concludes, with Marx, that history "has involved a progressive *objectification* and *externalization* of the productive forces, reaching its apotheosis in the industrial automaton" (ibid.: 311; emphasis in original). Disembedded from society, technology—like climate or ecology—is for most anthropologists something "whose study can be safely left to others. As climate is for meteorologists and ecology for ecologists, so technology is for engineers" (ibid.: 313). Ingold distinguishes technique from technology and asserts that "*there is no such thing as technology in pre-modern societies*" (ibid.: 314; emphasis in original). Technique is tacit, subjective, context-dependent, and practical, whereas technology is explicit, objective, context-independent, and—importantly—"can be transmitted by teaching in contexts *outside* those of its practical application" (ibid.: 316; emphasis in original). This line of reasoning is highly pertinent to the argument in this book, but we should note that Ingold's distinction between technique and technology is confined to the cognitive aspects of the two kinds of work processes, excluding their socio-ecological, metabolic requirements. If technological knowledge, unlike technique, can exist outside of its practical application, this again implies, as was demonstrated in the beginning of this chapter, that it is misleading to conceive of technology merely as a system of knowledge. It also requires access to an external energy source. In acknowledging this, furthermore, we have reason to doubt Ingold's assertion that technology occurs exclusively in modern societies. The engineering knowledge involved in transporting Queen Hatshepsut's obelisks from Aswan to Karnak could no doubt be taught in contexts outside of its practical application. The large-scale use of slavery, which Lewis Mumford (1967 [1934]) called the "mega-machine," was indeed a form of technology. In excluding its energy requirements, Ingold's analysis of the phenomenon of technology simultaneously excludes its relation to political economy.

Bateson's and Ingold's insights into the relational foundations of living things are very much in line with the perspectives of ecology and ecological systems theory, and both have extended these perspectives into other concerns such as those of psychology, cognition, and epistemology. Both advocate perspectives on communication that are closely related to those of Gestalt psychology and so-called field theory, which emphasize that a system generates qualitatively distinct phenomena which cannot be predicted from the aggregation of its parts. However, both these anthropologists appear to conceive of *social* systems as exclusively processes of communication, seemingly detached from the requirements of material metabolism. Although Bateson and Ingold are highly concerned with materiality and with transcending Cartesian dualism, they paradoxically remain constrained by a fundamentally Cartesian understanding of society as a nonmaterial system of communication. But if social relations are indeed a subset of ecological relations, as Ingold (2000) has proposed, we should expect them to be no less material than the flows of matter and energy which we identify as ecosystems. The perspectives of ecological economics and ecosemiotics are attempts to transcend Cartesian dualism from opposite directions, the former by showing that society is also material and the latter by arguing that nature is also communicative (Hornborg 2001a: 5). While the latter point is persuasively and extensively elaborated by Bateson and Ingold, the former appears to be outside their range of interest. Neither of these ecological systems theorists thus addresses the material metabolism and political economy of societal infrastructures such as globalized technologies. This may explain why neither of them has offered a "field theory" of technology.

Toward a Field Theory of Technology

To find a materialist theory of globalized *social* systems, we need to turn to the perspective of world-system analysis. The central argument of the world-system approach is that the specific trajectories of local, territorially bounded, or culturally homogeneous societies will largely be the result of much wider and less easily mapped fields of interaction. In Immanuel Wallerstein's (1974–1989) paradigmatic study, the economic development of individual nations in sixteenth-century Europe is explained in terms of an international division of labor within and beyond Europe. For Wallerstein, this sixteenth-century world-system comprised a "core," a "periphery," and an intermediate buffer zone or "semi-periphery." The core is the main locus of capital accumulation, based on an "unequal exchange" of its manufactured products for raw materials from its impoverished periphery. The sense in which such exchange is unequal has not been convincingly theorized in the world-system literature, but the issue is fundamental to this book (cf. also Hornborg 1998, 2001a, 2013).

Wallerstein's world-system model builds on Andre Gunder Frank's (1967) use of the concepts of "metropole" and "satellite" in dependency theory. Although both these theorists of global systems have been inspired by the historical materialism of Marxian concepts of imperialism, Frank's materialist understanding of global social systems is more literal. Frank (1998: 204; emphases in original) specifically attempts to reconceptualize technological development as "a *world economic process*, which took place in and because of the structure of the world economy/system itself." His conclusion is that "there was no *European* technology!" If capital accumulation is a global social process implicating the entire system, then so is technological development. This is closer to a "field theory" of technology. The maintenance of technological infrastructure in core areas of the world-system is tantamount to the metabolism of the global social order. Frank (2007) even considered the social and ecological disorder imposed by the expansion of Europe on its colonial periphery in terms of the displacement of entropy.

Taken together, the systems theory perspectives of Bateson, Ingold, Frank, and Wallerstein can help us transcend the epistemological inclination toward fetishism which Marx identified. Moreover, they can help us expand the Marxian notion of fetishism so as to refer not only to how we relate to money and commodities but also to how we relate to technology. Like biological organisms, technological objects are manifestations of intangible flows within wider webs of interactive relations. Technologies would not exist without these flows to sustain them. Just as DNA in itself would not suffice to generate an organism, blueprints cannot generate a technology. Both also require specific structures of exchange of matter and energy with their environments. The assemblages of artifacts that preoccupy Latour and ANT are indeed the substance of human sociality, but to endorse the attribution of autonomous agency to individual components of such assemblages, as ANT does, is precisely what Marx was trying to transcend. The challenge is to identify the existence and reproduction of technological infrastructures with the structures of exchange which are their foundation. However, to posit an equivalence between bounded material objects and intangible fields of communication contradicts the Cartesian habits of thought to which modern people have become accustomed.

The parallel between organisms and machines—biomass and technomass—can be extended further. Drawing on the natural laws of thermodynamics identified in physics, we can specify a basic requirement of the structures of exchange that maintain ordered processes such as living systems or the operation of machines. Biological organisms can stay alive only by importing more order-maintaining energy than they dissipate and discharge (Schrödinger 1967 [1944]). Highly ordered or "available" energy, referred to by physicists as "exergy" or occasionally "negative entropy," is what reproduces the internal order of any living system. The maintenance of this internal order, however, continuously dissipates the imported energy, making it less ordered and less available. In other words, the metabolic processes of living systems maintain order by exporting disorder, increasing entropy elsewhere. To stay alive, a living system must import more order than it exports. A basic requirement for its survival is thus an asymmetric flow of available energy, that is, a net import of order. This theoretical understanding of everyday metabolism is also applicable to technology.

Up until the Industrial Revolution, basically all the available energy that reproduced ecological and social systems on Earth derived from contemporary flows of sunlight. The asymmetric flows of available energy sustaining life simply consisted of the exchange of utilized solar energy for heat dissipated into space. Beginning in the late eighteenth century, however, the technological infrastructures that embodied the globalized social order increasingly relied on fossil fuels. As the physical requirements remained the same as ever, this meant that the fundamental condition for existence of the new societal infrastructures emerging in core regions of the world-system was an asymmetric exchange of energy and entropy not with outer space, but with other sectors of the world-system. In viewing such capital accumulation from the perspective of physics, we should be able to contribute to a more precise understanding of what Wallerstein and other world-system theorists have referred to as "unequal exchange."

As the remainder of this book emphasizes, the asymmetric exchange of physical resources in the world-system is intertwined in complex ways with the flows of money. The relation between the principles of physics and the discourse on economics is so complex and convoluted that most economists remain unconvinced that the economic growth of Europe since the late eighteenth century has built on unequal exchange with the rest of the world. However we represent this complex relation, we may know for certain that the expansion of technological infrastructure in core sectors of world society has represented a net import of physical resources and a net export of entropy. Not only can this conclusion be derived from the laws of thermodynamics and buttressed by a cursory glance at satellite images of nighttime lights, but it is also supported by statistics on global flows of matter as well as embodied energy, land, and labor (Lenzen et al. 2012, 2013; Yu et al. 2013; Simas et al. 2015). Moreover, the historical period since the dawn of the Industrial Revolution has seen a constant increase in anthropogenic emissions of carbon dioxide and is now often referred to as the "Anthropocene." The rising emissions of carbon dioxide from the combustion of fossil fuels is an incontrovertible illustration of the new kinds of entropy production associated with the growth of industrial technomass.

The main point in this chapter, however, is that the technological artifacts which surround us should be reconceptualized as embodiments of a highly unequal global social system. They are not intrinsically innocent inventions that in principle could be available to everyone, given sufficient purchasing power. Instead, we need to understand that the very existence of modern technology is a matter of global distribution. It is contingent on asymmetric flows of resources in the world-system. Purchasing power itself is increasingly skewed between different segments of world society. The global distribution of money, like technology, reflects the logic of a zero-sum game. The accumulation of technological infrastructure in core sectors of the world-system is contingent not on local ingenuity and entrepreneurship, but on differences in the market prices of labor and natural resources in different parts of the world.

CHAPTER 2

Land, Energy, and Value in the Technocene

On January 5, 1769, James Watt was granted a patent for a steam engine efficient enough to inaugurate the Industrial Revolution. It signified an epochal shift to fossil fuels as a source of mechanical energy. The technological ingenuity of the design of steam engines has been celebrated as the cause of this shift, but its prerequisites were fundamentally social. The British shift to steam power was a response to the world market demand for great volumes of inexpensive cotton cloth. Much of this demand came from slave traders in West Africa and slave owners in America, and these very slaves supplied the British cotton textile industry with inexpensive raw material. The steam engine, in other words, was made possible not only by James Watt's engineering, but by the eighteenth-century world-system in which capital accumulation in Britain was based on African slave labor and depopulated American land.

Fernand Braudel, a historian with an unusually wide-ranging and coherent view of the long-term continuities in human societies, endorsed an understanding of such continuities in terms of a succession of strategies for displacing workloads onto others. "There have always been," he writes, "a number of privileged persons (of various kinds) who have managed to heap on to other shoulders the wearisome tasks necessary for the life of all" (Braudel 1992: 65). We recognize, of course, the Marxian narrative leading from slavery through serfdom to wage labor. In this chapter, we shall introduce the basic idea which guides this book: that the phenomenon of "technological development" is in fact yet another strategy to be added to the list. We shall argue that modern technology, by and large, is not so much a replacement as a *displacement* of both work and environmental loads. The displacement of work is tantamount to an appropriation not only of energy and land, but also of human time. The finite time which humans have at their disposal is thus a scarce resource and has for millennia been the object of various elite strategies for redistribution. This means that political, economic, and technological power is ubiquitously intertwined with existential issues regarding the temporality of human life.

The shift to fossil fuels represented by Watt's steam engine inaugurated not only the Industrial Revolution, but also what is now being referred to as the Anthropocene, that is, a new geological epoch defined by the scope and intensity of anthropogenic modifications of global geochemistry. Human inventiveness is obviously a necessary condition for these developments, but, as evidenced by numerous experiments with steam power going back to the ancient civilizations of Egypt, Greece, and China, it is not a sufficient condition. To find the ultimate driving forces behind the Industrial Revolution, and thus also the Anthropocene, we must consider the organization of the eighteenth-century world economy.

The Metabolism of Societies

The ambition in this book is to consider modern technology as what Marcel Mauss called a total social phenomenon, viewed from the combined perspectives of history, sociology, economics, thermodynamics, ecology, epistemology, and culture theory. Anthropology is uniquely equipped to assemble such transdisciplinary perspectives on material aspects of contemporary societies, and to defamiliarize lifestyles and social arrangements which have come to appear natural and desirable. Those of us who enjoy the benefits of modern technology and patterns of energy consumption ought to recognize our lifestyle as the privilege of a global minority, and technology itself as a strategy for appropriating and redistributing time and space in global society. Such a reconceptualization of deeply rooted notions of technological progress and modernization would make it easier to grasp the nature of the global crisis that we are currently facing. Rather than fragment our understanding of this crisis into legitimate but separate worries over energy scarcity, environmental degradation, resource depletion, food shortages, climate change, global inequalities, and financial collapse, we need to realize that all these concerns are aspects of a single problem. This problem is the incongruous relation between modern social institutions and policies, on the one hand, and the Second Law of Thermodynamics,[1] on the other. The social arrangements and aspirations that are most fundamentally at odds with the Second Law of Thermodynamics are general-purpose money and beliefs in economic growth and technological progress. Of these illusions, the one that is most difficult to defamiliarize is undoubtedly that of technological progress.

Energy flows from the sun are the sine qua non of all living systems, including societies. Whereas organisms are programmed to harness such energy in specific and generally predetermined ways, human social systems have been able to devise a number of different strategies for accessing energy and distributing it among its members. As all human societies are organized in terms of more or less collective understandings of their own operation—the domain of cultural meanings—cultural images of energy and energy use constitute a formidable field for comparative study. Such study is inevitably contested and controversial, because access to and distribution of energy is everywhere closely connected to power. In fact, the very concept of "power" can be used to denote energy as well as social dominance.

For the 99 percent of its existence that our species has lived as mobile hunter-gatherers, humans have occupied specific niches in natural food chains, defined by their particular ways of extracting food energy from plants and animals. With the development of increasingly complex capacities for communication—involving more elaborate uses of language, symbols, and artifacts—human societies became more hierarchical, populations more concentrated and sedentary, and energy requirements more demanding. Beginning around 10,000 years ago, the domestication of plants and animals provided a more abundant and reliable energy niche for more complex societies in several parts of the world. The new demands on and sources of energy were recursively connected, so that, for instance, ceremonial feasting and chiefly generosity prompted expanded cultivation; more abundant harvests permitted larger settlements; the concentration of population demanded more intensified production; investments in productive facilities—generally farmland—required even greater concentrations of people for defensive purposes; and so on.

No doubt all these premodern societies, from hunter-gatherers to agricultural chiefdoms, had developed their own understandings of the energy flows that sustained them. Many of them seem to have recognized the sun as the source of vital power animating humans and the rest of the world. Agrarian empires were also ultimately dependent on the productivity of solar energy processed by plants, animals, and humans, and they, too, generally acknowledged—and in fact often worshipped—the sun.

From an abstract, comparative perspective, we can observe that these societies relied almost completely on the photosynthesis of various plant species and their conversion into food as well as on the mechanical work of animals and humans. What we have come to call land and labor were the ultimate energy resources, but they could be invested in capital in the form, for instance, of agricultural terraces, irrigation canals, roads, ships, armies, and temples. Capital is here defined as some kind of material infrastructure through which the extraction of energy can be increased.

Preindustrial agricultural cosmologies invariably recognized the productivity of the land as the foundation of human society. This was evident in

Europe as recently as among the eighteenth-century Physiocrats, and continues to be a dominant conception among nonindustrial agriculturalists in other areas of the world-system today (Gudeman 1986; Gudeman and Rivera 1990). In fact, even the physical energy of labor in these cosmologies is considered secondary to the "strength of the earth" (Gudeman and Rivera 1990). The labor theories of value of the nineteenth century were a product of industrial societies, for which the ultimate dependence on land had become increasingly opaque.

Money as the Condition for Technology

The mercantile empires that often handled long-distance trade between settled, agrarian polities developed a measure of power that had a much more tenuous connection to energy. Their niche was not land or labor, but exchange value, or purchasing power. This was particularly evident with the sixteenth-century emergence of transoceanic trading empires like the Portuguese, Dutch, and Spanish. If solar energy had been the vital force flowing through agrarian societies, money became the more abstract and elusive value that seemed to flow through and empower mercantile societies. The ambiguous relation between energy and money continues to elude us to this day. Purchasing power certainly appears to suffice to empower modern empires, but the crucial significance of energy for the economy does not seem to concern economics as a discipline and profession. Nor does a clear understanding of energy seem to be a part of the general public image of the organization of modern society.

But neither, of course, is there a clear understanding of money. The intellectual efforts that have been expended over the past two millennia to grasp the nature of money are impossible to summarize, and the general public today seems as baffled by its logic as ever. Suffice it to say that concepts of energy and money appear to fill similar functions in denoting a vital essence flowing through society. Like other species we are still, of course, as dependent on solar-derived food energy as ever, but the dominant cultural image of how modern society operates tends to marginalize such concerns in favor of a preoccupation with flows of money. This alienation from the vital flows which animate the biosphere derives in part from the historical experience of merchants and the social institution of money and in part from the nineteenth-century turn to inorganic, fossil energy, which was itself largely a consequence of the mercantile world order.

The concept of energy may seem as abstract and inaccessible as that of money, but it refers to objective, material flows that through various intuitive understandings have been part of human consciousness and rationality for hundreds of thousands of years. Its gradual replacement, over the past few millennia, by the concept of monetary value as the standard against which all things are assessed represents a cultural and ideological shift of momentous proportions. Unlike energy, money is fundamentally a fiction. It is not a physical condition of human existence, but a cultural convention the efficacy of which is contingent on human consciousness. When Marx observed that modern Europeans tend to conceive of money as an item that magically grows on its own account, comparable to the premodern worship of idols in West Africa, this was a quintessentially anthropological reflection in that it turned observations of exotic others back onto the observer's own familiar society. The economic reality in which modern humans are suspended is as culturally constituted as that of any other populations. This has been demonstrated by generations of economic anthropologists, but particularly convincingly by David Graeber (2011a). Money is truly a very peculiar idea and institution. It generates its own varieties of rationality that tend to be both imbued with and divorced from morality.

Numerous philosophers, social thinkers, and spiritual leaders have shown very persuasively that money is indeed a fetish. It is a reified representation of social exchange relations that in itself has no substance and no agency except through the ideas that people have about it. As such, it is the ideal tool for controlling people. The premises for rational transactions—for instance, commodity prices, interest rates, currency exchange rates—can be changed overnight without undermining basic trust in the rationality of money as such. Relative purchasing power can be redistributed in a population through adjustments of this or that regulation in ways so complex that it is impossible for anyone but the high priests of economics to decipher what is being done.

The notion that monetary exchange value is the substance or at least the driving force of society goes much further back than the Industrial Revolution, but it was a condition for it. There would have been no incentive for British textile manufacturers to radically intensify their production of cotton cloth if these commodities could not, by means of money, be exchanged for increasing volumes of embodied labor and land—for instance, in the form of imports of cotton fiber. The rationale of mechanization is intertwined with global differences in the prices of labor and resources. We seriously need to ask if industrialization would have occurred, if the African slaves harvesting cotton fiber on the colonial plantations had been paid standard British wages, and the owners of New World soils had received standard British land rent. The existence of modern technology, like the lucrative trade in spices, silver, or beaver pelts, is founded on strategies of conversion between different parts of the world market, where labor and land are very differently priced. This explains why the density of technological infrastructure continues to be very unevenly distributed over the face of the Earth, as can be observed on global satellite images of nighttime lights.

Thus money came to replace the concern of the eighteenth-century Physiocrats with fertile farmland as the basis of an affluent society. Although Thomas Malthus had worried about the availability of land as a constraint on economic growth, David Ricardo asserted that capital and labor could substitute for land, and Marx, too, was optimistic about the prospects of technological progress. These giants of economics appear not to have been very concerned about the fact that industrialization was fundamentally a strategy for Britain to appropriate, in terms of land area, an ecological footprint several times the size of its entire national territory, and, in terms of embodied labor, the toil of a workforce several times larger than its national population (Pomeranz 2000; Hornborg 2006, 2013).

The turn to fossil fuels as a source of mechanical energy was revolutionary in many ways. Geopolitically, it turned Britain into the most extensive empire the world has ever seen. In part, this was because its textile industry was able to oust its Indian competitors and thrive from the profitable triangular Atlantic trade which converted cotton cloth into African slaves, which were in turn converted into New World plantation produce, including cotton fiber. But fossil fuels also propelled the railways and the steamships with which Britain, frequently using military coercion, organized the metabolic flows of its global empire.

As already mentioned, fossil fuels also revolutionized economics and the public image of the economy. Up until the Industrial Revolution, energy requirements were basically synonymous with land requirements. The work of animals and humans always required land, either for animal fodder or for human food. This meant that there was a fundamental competition over land for production of food versus fodder, which farmers had been familiar with for millennia. Feeding draft animals such as horses and oxen claimed significant proportions of the agricultural landscape in preindustrial Europe. There was thus a limit on the amount of transport energy that was available, and on the distances that bulk goods such as food or fodder could be transported, before the quantity of energy used to transport the goods exceeded the energy content of the goods themselves. This constraint was a consequence of the fact that both kinds of energy represented the product of a quantity of land.

Fossil fuels provided a form of energy that did not compete with food production or other uses of the land. This meant that access to land no longer represented the ultimate constraint, as it had to the Physiocrats and to Malthus. Provided that the price of fossil fuels was low enough, it did not matter if the energy expended in transports exceeded the energy content of the cargo. From now on, the same logic applied to production, including agriculture. Relative market prices of various forms of energy, including labor, determined input–output ratios and the feasibility of different kinds of technology. Industrialized production and mechanized transports—of imports as well as exports—went hand in hand. The economic expansion of Britain was determined by the market prices of cotton textiles, slaves, and coal, not by the ability of British crops to harness solar energy.

It is not difficult to imagine how this fundamental transformation of economic rationality must have impacted on human perceptions of society. Natural constraints were no longer absolute but could be transcended with the help of new technology. If British soils had been exhausted of nutrients, they could be replenished through the import of guano and phosphates from islands in the Pacific. The extent to which this relied on slave-like working conditions on those islands as well as in the British coal mines was made more or less invisible by the impersonal logic of the market, as were the ecological consequences (Clark and Foster 2012). The concept of technology from now on signified the seemingly magic capacity of some humans to improve their conditions through sheer ingenuity. Technology thus continued to be perceived as more or less completely a product of inventiveness, without regard to the particular kinds of global exchange relations on which it depended. Moreover, technology was perceived as inevitably progressing toward higher and higher efficiency. But, like rationality, efficiency is ubiquitously defined by the cultural and societal context. If conceived in terms of an input–output analysis, the parameters for assessing efficiency were not related to expenditures of energy, but rather to the input and output of money, and of upper-class human time.

It may be helpful, at this point, to add a reflection on the classic conceptualization of unequal exchange by Arghiri Emmanuel (1972). In a nutshell, he argued that, because of international differences in wages, poor nations are obliged to export greater volumes of embodied labor than they would do if wages were uniform. If we exclude Emmanuel's deliberations on labor "value" (see below), this is a perfectly valid observation. International wage differences generate asymmetric flows of embodied labor time, the appropriation of which contributes to underdevelopment in the periphery. But let us also consider this analysis from the converse perspective. If technological progress such as the Industrial Revolution is understood as a process of capital accumulation in the core, at the receiving end of a relation of unequal exchange, it is also a product of international differences in wages. It, too, would not occur if wages were uniform. Needless to say, this conclusion should pose certain problems for those orthodox versions of Marxism which celebrate the "inexorable" progress of the productive forces.

Technology as the Displacement of Slavery

The relation between modern, predominantly fossil-fuel technology and slavery is complex. The colonial slave plantations that supplied the British textile industry with cotton fiber were obviously part of the conditions for industrialization in the first place (Inikori 1989, 2002). Ironically, however, industrialization appears to have been an important factor in the official abolition of slavery (Mouhot 2011). Today it is common to think of the access to modern technology in terms of its equivalent number of "energy slaves." This is actually much more than a metaphor.

Part of the legacy from ancient Greece and Rome was the delegation of work to other beings that were more or less degraded to things. The idea of externalizing toil from the bodies of free men and delegating it to purportedly mindless agents was fundamental to these civilizations of antiquity. Faced with increasing difficulties in procuring slaves, the choice between doing the work yourself and devising new mechanical contraptions seems predictable. In order to maintain a traditional lifestyle and identity, landowners in the fifth century AD tended to replace slaves with the first machines, that is, water mills (Debeir et al. 1991: 39). The rationality of such technological progress then as now hinges on the relative prices of labor and resources. The logic of having a water mill built instead of purchasing slaves is essentially the same as using a vacuum cleaner instead of hiring a housemaid. In both cases, we could add, the owners of technology are able to imagine that technological progress has done away with degrading, low-wage toil. In both cases, however, a closer familiarity with the socioeconomic conditions under which the new technology is manufactured and maintained might have given them a different perspective. To take the example closest at hand, it is far from evident that the modern employees of Chinese vacuum cleaner manufacturers are better off than European or American housemaids were a century ago. Middle-class households in Europe and the United States have substituted domestic appliances of various kinds for their housemaids, but the Chinese factory workers who now produce these appliances are no more affluent than the low-wage Europeans and Americans whose labor those machines replaced a few generations ago.

From the perspective of privileged sectors of society, investment in new technology is understandably perceived as progress. This conviction has for at least two centuries been fundamental to dominant conceptions of history, development, and modernization. But to a large extent, technological progress has been the privilege of affluent elites, and the very existence of the new technology has relied on the appropriation of embodied labor and resources from an increasingly impoverished periphery. The investments in steam technology in early nineteenth-century Britain were indissolubly connected to the Atlantic slave trade and the cotton plantations in the American South. They relied on a continuous, unequal exchange of embodied labor and land between the industrial core and its colonial periphery.

World society remains highly polarized between high-tech core areas with high levels of per capita purchasing power and energy consumption, on one hand, and peripheral areas with much lower levels of purchasing power and energy consumption, on the other. The unequal exchange of embodied labor time in the modern world was demonstrated 40 years ago by Emmanuel (1972) and has recently been confirmed by Simas et al. (2015).[2] The unequal appropriation of embodied land has been amply documented by the research on ecological footprints (Wackernagel and Rees 1996) and on the asymmetric global flows of materials, embodied land, and embodied energy (Lenzen et al. 2012, 2013; Yu et al. 2013; cf. Dorninger and Hornborg 2015).[3]

A relevant question for social scientists, including anthropologists, to ask at this point in world history is whether modern technology has really replaced slavery, or merely *dis*placed it. If the concept of slavery is defined not primarily in terms of physical violence, but more fundamentally in terms of being coerced to perform alienating, low-status tasks for the benefit of a privileged elite, a significant part of the world's population would qualify as slaves. Seemingly neutral concepts such as technology and the world market organize the transfer of their embodied labor and resources to an affluent minority. From this perspective, the operation of technology represents the deflected agency—the labor energy—of uncounted millions of laborers, harnessed for the service of a global elite. To view technology in terms of a set of energy slaves is thus indeed more than a metaphor.

"Value" as Mystification

The most serious criticism of mainstream economics and what Aristotle called chrematistics—the art of managing money, as opposed to real resources—has come out of Marxist and ecological economics (Martinez-Alier 1987). These two schools of thought have actually converged in their criticism, demonstrating how exchange rates set by market prices can conceal an asymmetric exchange of labor or resources that are significant for macro-level processes of development versus underdevelopment. It is interesting to note that both schools have in fact been concerned with the way in which monetary prices mystify flows of energy through society. The net flows of embodied labor-power emphasized by Marxists are no less a form of energy than the flows of resources (such as *emergy*, originally conceived as shorthand for embodied energy) studied by ecological economists (Odum 1996). Some theorists have explicitly compared these two approaches to unequal exchange (Lonergan 1988). Analytically, the arguments are indeed identical. Unequal exchange is posited to occur when some kind of "value" (labor value or energy value, respectively) is being underpaid.

Predictably, such arguments will not convince mainstream economists, and in this particular respect I must agree with them. As we shall see in chapter 5, the problem is couching the discussion in the idiom of value. Anthropologists are well aware that value is culturally constituted and cannot be derived from Marxist theory or from physics (cf. Sahlins 1976; Baudrillard 1981 [1972]; Bourdieu 1984 [1979]). To suggest that Marxist or ecological economists have a better understanding of what is valuable than market actors, and that the latter consistently underpay these more essential values, is the wrong way to approach the problem. The major mistake that these theorists make is to use the concept of value for some kind of material flow that is not in itself the object of valuation. It is not the quantity of embodied labor or energy that determines how much a consumer is willing to pay for a given commodity. The surplus value that provides profits for capitalists is not a metaphysical product of labor-power—one of several possible sources of energy—but simply the difference between the cost of inputs and the gain from sales of the output. In addition to human labor, the inputs may include, for instance, the work of draft animals, fuels, and raw materials.

The paradox, then, is that critics of mainstream economics, in struggling to expose the ideological function of the market in mystifying asymmetric flows and to identify various forms of energy as the asymmetric flows thus mystified, have resorted to the mercantile notion of value to underpin their argument. This notion has for centuries pertained to money (i.e., exchange value) and consumer preferences. It belongs to the vocabulary of the market and should not be confused with the objective, material flows that both Marxist and ecological economists are concerned with. In order to argue that the world market conceals asymmetric flows of energy that contribute to global inequalities in the distribution of technology, purchasing power, and environmental quality, we would do best to talk about precisely that: asymmetric flows of *energy*.

In thus distinguishing material flows of energy from semiotic flows of money values, while simultaneously relating them to each other, we would be presenting an argument that would not be easy to dismiss. To observe that the accumulation of technological infrastructure in certain areas of the world, visible on satellite images of nighttime lights, would be impossible without a net input of available energy is simply based on the Second Law of Thermodynamics. Similarly, to observe that energy is dissipated in economic processes, implying that industrial products contain less available energy than the fuels and raw materials that were used in making them, is also completely in line with the law of entropy (Georgescu-Roegen 1971). From a physical perspective, in other words, production is destruction. The creation of consumer value or utility is simultaneously the creation of entropy. Finished products must be priced higher than the inputs—labor, fuels, and raw materials—but inexorably represent less available energy. The dissipation of resources is thus blindly and continuously rewarded by the market with more resources to dissipate. This in turn means that increasing quantities of energy and materials will be accumulated in what Stephen Bunker (1985) called the productive sectors of the economy, while the extractive sectors are increasingly impoverished. The argument thus accounts well for the polarizing tendencies of the world economy, without resorting to any normative, contestable propositions. Unequal exchange in this sense simply means asymmetric. Whether it is also unfair is a moral conclusion that is up to the reader, rather than an assumption of the argument as such.

But, someone might object, even if embodied labor or energy is not a value but a physical measure, wouldn't it be valid to propose that it is being underpaid? Couldn't the inequalities in the world be leveled out by adjusting exchange rates? Unfortunately, the basic problem is not simply that prices of embodied labor, energy, and resources are too low in relation to those of finished goods, but that if they were higher, there would be no incentive to continue the exchange. What would be the point of mechanization if it was not profitable? As argued above, the rationality underlying British industrialization in the nineteenth century was founded on the much lower price of land and labor in the colonies. The foremost rationale of industrial capitalism is to *not* have to pay for the costs of increasing social and ecological disorder in the surrounding world.[4] This logic continues to pervade technological progress to this day.

Another objection, among anthropologists, might be that we are not physicists but social scientists. We can only be concerned with cultural constructions, the argument might go, not with putatively objective conditions. But I would counter that it is precisely in its transdisciplinarity that anthropology has its greatest potential, as mediator between social and natural science. In an unforgettable sentence, Roy Rappaport (1994: 154) summarized the predicament of our field: "[A]ny radical separation of [the objective and the subjective] is misguided, not only because meanings are often causal and causes are often meaningful but because, more fundamentally, the relationship between them, in all its difficulty, tension, and ambiguity, expresses the condition of a species that lives, and can only live, in terms of meanings it itself must construct in a world devoid of intrinsic meaning but subject to natural law." To "comprehend the fullness of its subject matter's condition" (ibid.), anthropology must not only remind the natural scientists that humans live in terms of meanings, but also remind the social scientists—including economists—that societies are subject to natural law.

The Anthropology of Sustainability in the Anthropocene

A defining feature of an anthropological perspective is that it acknowledges the importance of *cultural specificity*. As Marshall Sahlins (1976: 170) succinctly put it, "no object, no thing, has being or movement in human society except by the significance men can give it." Any discussion of human–environmental relations, patterns of consumption, power structures, or economic worldviews would be incomplete without considering the particular systems of meaning that organize them. Yet very little mainstream discourse on sustainability, environmental problems, or the economy is concerned with cultural aspects. Natural scientists, technologists, and economists have to a large extent monopolized these discussions, leaving anthropologists with a sense of dismay at finding publicly accepted problems and their solutions so narrowly and simplistically defined.

Anthropological, cultural analyses should have crucial things to say about past, present, and future concerns with sustainability, yet remain a very marginal perspective. Instead, a growing public concern about the prospects of socioecological collapse a few years back provided an ornithologist—a biogeographer—with the opportunity to produce a best seller on the turbulent history of human societies (Diamond 2005). To a large extent, the marginalization of anthropological perspectives on sustainability probably derives from the fact that reflexive, cultural approaches are intrinsically more difficult to grasp—and above all, to *apply*—than ecological, technological, or economic ones. But there are doubtlessly also ways in which anthropologists could exert themselves a bit more to become relevant.

The greatest problem derives from the tenacious separation of social and natural science. Those researchers who seem to be the most concerned about the future of the biosphere—the natural scientists—generally have the bluntest analytical tools for understanding the causes of anthropogenic environmental degradation, while those who possess such tools—social scientists, including anthropologists—generally tend to be less concerned with the biosphere as an objective, biophysical reality. In having been trained or even required to approach the environment with quotation marks, as a contested or negotiated cultural construction, anthropologists often seem at a loss when expected to say something about the *real* environment beyond human perceptions. Certainly, natural scientists need to realize that cultural sign systems such as language, money, and power are components of ecosystems, organizing significant aspects of their flows of matter and energy, but social scientists conversely need to realize that flows of matter and energy are fundamental to social systems, and need to be taken into account in any explanation of development, underdevelopment, and collapse. Even in as holistic a field as anthropology, it is striking how the most incisive discourses on political economy, economic anthropology, and cultural aspects of human exchange (for instance, Graeber 2001, 2011a, 2011b) generally have so little to say on the material substance of exchange—for instance, on energy, ecology, and technology— and how even discourses on materiality tend to avoid questions of *materially* asymmetric global resource flows. Anthropologists—at least the four-field

Land, Energy, and Value in the Technocene • 29

kind—should be uniquely equipped to think in truly transdisciplinary ways about how material and semiotic processes are intertwined.

This is why anthropologists should also be able to contribute significantly to the currently unfolding discourse on the Anthropocene. This discourse represents a convergence of Earth system natural science and what I will refer to as post-Cartesian social science. Both fields suggest that the Enlightenment distinction between nature and society is obsolete. Now that humanity is recognized as a geological force, the story goes, we must rethink not only the relations between natural and social sciences but also history, modernity, and the very idea of the human. Indeed, the increasingly inextricable interfusion of nature and human society is incontrovertible, as evidenced not only by climate change but also by several other kinds of anthropogenic transformations of ecosystems.

For decades, having believed these circumstances to be self-evident, some anthropologists may be surprised by the intensity and also the character of the philosophical import that is currently attributed to them. The theoretical implications of the interfusion of nature and society, and the imperative of transdisciplinary approaches to human–environmental relations, were already prominent in social-science agendas in the 1990s (for instance, Haraway 1991; Narain and Agarwal 1991; Croll and Parkin 1992; Latour 1993; Descola and Pálsson 1996; Peet and Watts 1996; Escobar 1999). Fields such as environmental anthropology, political ecology, development studies, and science and technology studies (STS) were attempting to deconstruct the nature/society distinction more than 20 years ago.[5] Rather than embroil ourselves in increasingly obscure deliberations on the possible philosophical implications of this shift, it should now be incumbent on social scientists to try to be as clear as possible about the societal and not least political issues that it raises.

The questions we need to ask are: In what sense should the idea of the Anthropocene change our understanding of human–environmental relations, history, and modernity? If post-Cartesian perspectives can help us grasp climate change, how can they simultaneously illuminate the history of technology and development? Do they imply a complete dissolution of the categories of nature and society, or merely their reconceptualization? Is the notion of the Anthropocene an adequate designation for the current period? What are the prospects for humanity surviving the planetary transformations that it has set in motion?

Can We Dispense with the Categories of Nature and Society?

Let us initially emphasize that the physical mixing of nature and society does not warrant the abandonment of their analytical distinction. Rather, precisely this increasing recognition of the potency of social relations of power to transform the very conditions of human existence should justify a more profound engagement with social and cultural theory. It is deeply paradoxical and disturbing that the growing acknowledgment of the impact of societal forces on the biosphere should be couched in terms of a narrative so dominated by natural sciences such as climatology and geology.

A prominent role of science seems to be to represent technological progress as natural, as if capitalist expansion was founded exclusively on innovative discoveries of the nature of things, and as if the social organization of exchange had nothing to do with it. Constrained by our Cartesian categories, we are prompted by the materiality of technology to classify it as belonging to nature rather than to society. The post-Cartesian solution to this predicament would be to abandon the categories of nature and society altogether. But to acknowledge that nature and society are inextricably intertwined all around us—in our bodies, our landscapes, our technologies—does not give us reason to abandon an analytical distinction between aspects or factors deriving from the organization of human society, on the one hand, and those deriving from principles and regularities intrinsic to the prehuman universe, on the other. For example, the future of fossil-fuel capitalism no doubt hinges on the relation between the market price of oil and the Second Law of Thermodynamics, but I cannot imagine that we have anything to gain from dissolving the analytical distinction between the logic of the world market and the laws of thermodynamics.

Regardless of how we represent them, the laws of thermodynamics have been in operation as long as there has been a universe, billions of years before the origins of human societies. They are an undeniably natural aspect of human existence that pervades everything we do, and yet have not been, and cannot be, the least altered by human activity. In contrast, modes of human social organization such as markets are ephemeral constructs that can be fundamentally transformed by political decisions or the vicissitudes of history. Yes, thermodynamics and markets are intertwined in fossil-fuel capitalism, but this is no reason to deny that the former belongs to nature and the latter to society.

In similar ways, it is possible *in principle* to trace the interaction of factors deriving from nature and society. It is feasible, for instance, to estimate what the concentration of carbon dioxide in the atmosphere would have been today, if the additions deriving from human social processes had not occurred. Human societies have transformed planetary carbon cycles, but not the carbon atoms themselves. If the categories of nature and society are obsolete, as it is currently fashionable to propose, this only applies to images of nature and society as bounded, distinct realms of reality. At the risk of being unfair to René Descartes, I follow the convention of referring to such distinctions as examples of Cartesian dualism. It seems trivial to observe that such bounded, distinct realms do not exist (who would object today?), but it remains justified to identify the logic of natural and societal phenomena separately, prior to demonstrating how they interact in practice. The challenge of transdisciplinarity is not to jettison intradisciplinary expertise, but to acknowledge that several kinds of specialized expertise may be required to understand socioecological processes. Such are the difficult but crucial ambitions of transdisciplinary fields like ecological economics and political ecology. The disciplines of both physics and economics, for instance, need their devoted scholars, but it would be mistaken to expect either of them in isolation to provide a full account of fossil-fuel capitalism.

In consequence with the abandonment of Cartesian dualism in our approach to anthropogenic transformations of the biosphere, we have no less reason to reconsider human economies and technologies as similarly hybrid phenomena interlacing biophysical resources, cultural perceptions, and global power structures. Such insights deserve to be pursued not only at the microlevel of our interaction as individuals with specific artifacts, as advocated by ANT, but more importantly at the macrolevel, where the global assemblages of artifacts that I have called technomass (Hornborg 2001a) indeed are the very stuff of a highly inequitable world-system.[6] It is in this global sense that the social dimensions of technology are the most interesting. By viewing it as a system-wide totality, we can detect how global power relations are delegated to, and buttressed by, technology. Now that we are addressing the environmental predicaments of the Anthropocene from a truly global perspective, why should we not look at the sociotechnical networks that brought us here in the same way?

Conventional historiography depicts the Industrial Revolution as the product of British ingenuity and as a contribution destined to diffuse among all humankind. A scrutiny of the transition to fossil fuels in late eighteenth-century Britain, however, reveals the extent to which the historical origins of anthropogenic climate change were predicated on highly inequitable global processes from the start. As we have shown, the rationale for investing in steam technology at this time was geared to the opportunities provided by the constellation of a largely depopulated New World, Afro-American slavery, the exploitation of British labor in factories and mines, and the global demand for inexpensive cotton cloth. It would thus be highly misleading to conceive of the *anthropos* starring in the Anthropocene narrative as the human species (Malm and Hornborg 2014). Humanity as a collective has never been an agent of history, and the technological fruits of the Industrial Revolution continue to be very unevenly accessible to different segments of world society. This uneven distribution of modern, fossil-fuel technology is in fact a condition for its very existence. The promises it held out to humanity were illusory all along: the affluence of high-tech modernity cannot be universalized, because it is predicated on a global division of labor that is geared precisely to huge price and wage differences between populations. What we have understood as technological innovation is an index of unequal exchange.

Let us recapitulate this by properly defining the notion of "modern technology." The conditions of technological innovation were radically transformed in the late eighteenth century. We usually think that the decisive factors were engineering science and the adoption of fossil fuels, but none of this would have been possible without the global social processes which made the relative prices of labor and resources on the world market *prerequisite* to technological progress in Europe. Up until that historical point, technology was founded on local ingenuity, and understood as such. Beyond that point, and for over 200 years now, the understanding of technology as founded on mere ingenuity has persisted, but has become highly inadequate. Ingenuity is a necessary but not *sufficient* condition for modern technological progress. Global price relations are systematically excluded from our definition of technology, even though, by organizing asymmetric resource flows, they are crucial for its very existence. Much as inexpensive labor and land in colonial cotton plantations were fundamental to the Industrial Revolution, it today remains essential for high-tech society that prices of oil and other resources are manageable. What we have thought of as the history of human inventions is actually the history of rising inequalities within an increasingly globalized economy. When Paul Crutzen (2002: 23) refers to "James Watt's design of the steam engine in 1784," evoking our conventional understanding of an ingenious but seemingly random technological breakthrough, neither he nor his readers will be inclined to reflect on the extent to which this invention implicated colonialism and slavery.

The density of distribution of technologies that are ultimately dependent on fossil fuels by and large coincides with that of purchasing power. These technologies are an index of capital accumulation, privileged resource consumption, and the displacement of both work and environmental loads. After more than 200 years, we still tend to imagine technological progress as nothing but the magic wand of ingenuity that, with no necessary political or moral implications elsewhere, will solve our local problems of sustainability. Universities throughout the world reproduce this illusion by entrenching the academic division of labor between faculties of engineering and faculties of economics. But globalized technological systems essentially represent an unequal exchange of embodied labor and land in the world-system. The worldview of modern economics, the emergence of which accompanied the Industrial Revolution in the hub of the British Empire, systematically obscures the asymmetric exchange of biophysical resources on which industrialization rests. This disjunction between exchange values and physics is as much a condition for modern technology as engineering.

Is the Notion of the Anthropocene Adequate?

The uneven accumulation of technomass visible on satellite photos of nighttime lights proceeds by means of a simple algorithm: the more fossil fuels and other resources it has dissipated today, the more it will afford to dissipate tomorrow. This account of our entry into the Anthropocene does not refer to the biological properties of the species *Homo sapiens*, but to a specific form of social organization that emerged very recently in human history, as a strategy of one segment of humanity to dominate the remainder. This form of social organization continues to be propelled by the interests not of our species, but of a social category (Malm and Hornborg 2014). As of 2008, less than 20 percent of the world's population was responsible for over 70 percent of carbon dioxide emissions since 1850 (Roberts and Parks 2007). An average American today emits as much carbon dioxide as 500 average citizens of some nations in Africa and Asia (ibid.). It must thus be the work of social science to identify the drivers of rising emissions.

The dominant Anthropocene narrative, of course, does recognize that climate change derives from human activities, but these activities are then viewed as expressions of innate traits of our species. Rather than examine their societal and political drivers as factors that can be transformed, the narrative tends to represent them as natural and inevitable features of our biology. But phenomena such as worldviews, property relations, and power structures are *social* phenomena. They are beyond the horizons of natural science, because they require analytical tools that natural scientists are not provided with.

This is not to deny that human organisms are uniquely equipped to develop capitalism. Our semiotic capacity for abstract representation and language, which had enormous survival value for hundreds of thousands of years of hunting and gathering, finally generated general-purpose money and the globalized economy, which in turn made the Industrial Revolution feasible. The world-systemic events of the eighteenth century were products of a global history of increasing interconnectedness and inequality ultimately founded on the human capacity for abstract representation. The big question is whether this capacity will be of any use in redesigning our global economy for survival. To challenge the species-centrism of the Anthropocene narrative is to make two important points that are often disregarded by natural scientists: (1) the incentives, benefits, and negative repercussions of industrialization are very unevenly distributed among social categories within the human species and (2) there is nothing biologically inevitable about the institutions and forms of social organization that we know as capitalism.

Dipesh Chakrabarty (2009) correctly observes that we now need to integrate the history of our species with the history of capital, but he does not provide us with any feasible suggestions on how to proceed. He is completely silent on how our biological capacity for abstract representation—as in language and other semiotic systems—is prerequisite to the very idea of money, and how money was in turn prerequisite to the Industrial Revolution that inaugurated the Anthropocene. But it is precisely through this chain of events that studies of natural and human history, while each reserving its specific arsenal of concepts and methods, can be integrated. Modern technology is the pivot of both, because it implicates both biophysical and sociocultural dimensions of our increasingly globalized history. Rather than imply that climate change is the inexorable consequence of the emergence of *Homo sapiens*, as suggested by the notion of the Anthropocene, it would be better if the geological epoch inaugurated in the late eighteenth century be named the *Technocene*.[7]

It is disturbing that social scientists often seem to be retreating from the playing field defined by Earth system science. Whether intended or not, this is a widespread consequence of the assertion that the distinction between natural and social is obsolete (Latour 1993). This dismal verdict on centuries of social science appears to be geared to the conviction of ANT that there is no difference between the agency of human beings equipped with perceptions and intentions, on the one hand, and that of rocks, artifacts, and other nonliving things, on the other. This foundational assumption of ANT is fundamental also to the approach of Latour (2013) and his followers to the Anthropocene. But if humans and their artifacts can be shown to be "actants" of very different kinds, as is one of the aims of this book (see particularly chapter 6), it might help us retain some of our faith in social science. The social sciences, and prominently among them anthropology, have long struggled to respectfully assimilate perspectives from natural science. An example is the field of political ecology.

Political Ecology in the Technocene

The discourse on political ecology emerged in the early 1970s as an ambition within several social sciences to relate local ecological dilemmas, primarily in what was then known as the Third World, to global political economy. Over the decades, two main lineages of research can be discerned: one acknowledging an objective nature and a set of actors contesting each other's claims to resources, the other inspired by poststructuralist theory to deconstruct images of nature as well as the identities and claims of the actors (Escobar 1999). Disarmed by its own relativism, the latter, constructivist approach has predictably yielded fewer substantive challenges to capitalist extractivism. Pursuing constructivism ever deeper into philosophical opacity, the "political ecology" of Bruno Latour (2004) has been criticized for even more radically disarming political criticism (e.g., Söderberg and Netzén 2010; Wilding 2010; Hornborg 2014a).[8] The emphasis of the constructivist wing of STS on microsociological case studies of individual actors has gone hand in hand with a rebuttal of more inclusive socioeconomic power structures like capitalism or even society (Söderberg and Netzén 2010: 100–102).

We must insist, however, that only *societies*—organized assemblages of interacting human beings—negotiate meanings, generate relations of unequal exchange, and enable people to exert power over each other. Of course, all these social relations are stabilized through the recruitment of nonhuman components into their networks, and of course they are to a large measure shaped by the specific features of these nonhuman components, but the driving forces and the glue that reproduce them are irreducibly *social* in the sense that they hinge on the incentives, intentions, and agency of interacting human subjects.

All this certainly does not mean that social power does not have material components. On the contrary, it always does. Our challenge as social scientists is to show how these material dimensions of power are systematically obscured in hegemonic discourses and worldviews—for example, how the unequal global exchange of labor energy and other biophysical resources is obscured in mainstream economics, how this unequal exchange is prerequisite to our obsession with technological progress, and how technological progress is thus ultimately a fetishized account of the global displacement of work and environmental burdens to social categories with less purchasing power.

The Anthropocene narrative is rapidly gaining ground as our hegemonic discourse and worldview. The question is how we relate to it as social scientists. To the extent that it prepares us to acknowledge the constant interlacing of nature and society we can only hope that this vision will not be confined to the study of our changing biosphere and atmosphere, while we remain blind to the interlacing of the material and the social in our globalized technologies. A post-Cartesian understanding of the Industrial Revolution should fundamentally reframe the discourse on political ecology. Rather than dream of advanced technological solutions to problems of ecological sustainability,[9] we would recognize most modern technologies as social strategies for *displacing* problems—labor as well as environmental loads—to areas where labor and environmental degradation are less expensive. Instead of technological utopianism, this radical reconceptualization of technology should prompt us to critically consider the role of general-purpose money in orchestrating asymmetric transfers of labor-power and natural resources in the world-system.

Given this analysis of the respective roles of capitalism and the species-specific characteristics of *Homo sapiens*, what kind of visions for a sustainable future can it support? Let us first establish that there is no inevitable connection between human biology and industrial capitalism.[10] Our capacity for abstract representation was prerequisite to capitalism, but only by means of the specific sociocultural institution of general-purpose money. It was through the globalized circulation of general-purpose money that all the ingredients of the Industrial Revolution—American fields, African slaves, cotton fiber, British workers, coal, and cotton textiles—were transformed into commensurable and interchangeable commodities. The generalized commoditization of all this human time and natural space, which made industrialization possible, is *not* an inexorable consequence of the human capacity for representation. If the economic strategies generating globalization and industrialization are root causes of the perilous prospects of climate change, it should be theoretically possible to avert this threat by modifying the conditions of economic rationality. It should be feasible, in principle, to organize a monetary system that restricts the interchangeability of products to specific spheres of exchange through the use of special-purpose currencies. This is not to suggest that certain kinds of exchanges should be prohibited, but that the options available to individual actors should encourage transactions that substantially reduce the consumption of fossil fuels and other practices contributing to environmental degradation. This is the topic of chapter 8.

CHAPTER 3

The Magic of Money

In the late thirteenth century, the Venetian merchant Marco Polo reported with some astonishment that the government of the Chinese Empire issued paper money made from the bark of mulberry trees, which was considered as valuable as the metals, gems, and pearls for which it was exchanged. Polo seems to have interpreted the paper currency as a conspiracy, enabling the emperor Kublai Khan to accumulate valuables at a negligible cost. The Khan, Polo remarked, "buys such a quantity of those precious things every year that his treasure is endless, whilst all the time the money he pays away costs him nothing at all" (Yule 1871:380).

The idea of issuing paper money to replace cumbersome metal coins was soon adopted in Renaissance Italy (Weatherford 1997). It represents a pivotal step in the semiotic transformations of money which continue to this day. Polo's comment illustrates that, at one level, the users of paper money maintain a conviction that the substance of this currency is not as intrinsically valuable as the substance of metal coins. Yet he assures his readers that merchants in China happily accepted payment in paper money. While thirteenth-century merchants no doubt distinguished between what they perceived as the authentic value of gold and silver, on the one hand, and the acknowledged but basically fictive value of paper money, on the other, a semiotic perspective allows us to conclude that, whether gold or paper, money is merely a socially accepted token of exchange value. This means that the substance of the money token is actually arbitrary, as long as it is in limited supply and cannot be easily counterfeited. From this perspective, the main difference between precious metals and paper money is that limited supply in the former case is a consequence of the relatively sparse natural occurrence of gold and silver, whereas in the latter case scarcity is artificially produced. Also, counterfeiting paper money is more feasible than alchemy.

The shift from precious metals to paper in retrospect clarifies that artifacts serving as money tokens are no more than representations of abstract exchange value—they are thus ultimately coveted for their potential use in social transactions, not for some imagined, essential value intrinsic to the money tokens themselves. If it were not for international agreements such as those of Bretton Woods, gold could conceivably be as useless a medium of exchange in some cultural contexts as seashells are to modern Europeans.

This understanding of money, however, simultaneously implies that there is no such thing as intrinsic value. If value ubiquitously pertains to social relations, any notion of intrinsic value is an illusion.[1] Although the European plundering and hoarding of gold and silver, like the Melanesian preoccupation with *kula* and the Andean reverence for *Spondylus*, has certainly been founded on such essentialist conceptions of value, the recent representation of exchange value in the form of electronic digits on computer screens is a logical trajectory of the kind of transformation propagated by Polo. It is difficult to imagine how money appearing as electronic information could be perceived as possessing intrinsic value. This suggests that electronic money, although currently maligned as the root of financial crisis, could potentially help us rid ourselves of money fetishism. Paradoxically, the progressive detachment of money from matter, obvious in the transitions from metals through paper to electronics, is simultaneously a source of critique and a source of hope.

As countless philosophers and social thinkers over the centuries have recognized, the phenomenon of money is recursively intertwined with central features of the human condition, from modes of cognition, religion, and morality to power, exploitation, warfare, and the nation state. It is also foundational to the sociological condition of modernity, frequently characterized in terms of inclinations toward abstraction, interchangeability, individualism, and alienation. The very concept of money is thus a pivotal cultural phenomenon that ought to be at the center of anthropological deliberations on modernity, development, and sustainability. Indeed, it can be argued that the concept of "indigenous people," which is often a central concern of anthropological research, could be defined as those segments of humanity who—in theory or in reality—have not yet succumbed to an outlook conditioned by modern money. Indigenous people fascinate modern people because they represent an imagined alternative—and a resistance—to money.

The Semiotics of Money

In view of the extent to which market economies, capitalism, and the conceptual framework of conventional economics are founded on the logic of money, it is appropriate to present some general reflections on this unique semiotic phenomenon. Semiotics (from Greek *semeion* = sign) is the study of sign systems. A semiotic perspective on money would thus approach it as a kind of sign, comparable to other systems of signs such as language, gestures, clothing, and so on. Signs are means of communication that presuppose subjects, meanings, codes, and interpretations. They are by no means restricted to the human species but seem to be pervasive in living systems at all levels of scale, from the internal biochemistry of individual organisms (Sebeok and Umiker-Sebeok 1992; Hoffmeyer 1996) to the various kinds of communication between the myriad organisms of an ecosystem (von Uexküll 1982 [1940]; Nöth 1998; Hornborg 2001b; Kohn 2013). The analytical study of sign systems was pioneered by Ferdinand de Saussure (1916) and Charles Sanders Peirce (1931–1958), but the strong linguistic focus of the former has not been conducive to wider comparative approaches such as those inspired by Peirce.

General-purpose money is a peculiar kind of sign. It seems impossible to classify it as belonging to one of Peirce's three general categories of signs: index, icon, or symbol. The distinction between these three types of signs is based on differences between how they relate to their referents (i.e., what they refer to): an *index* relates to its referent through contiguity, an *icon* through similarity, and a *symbol* through convention. A money sign, whether a coin, a paper bill, a check, or an electronic digit, does not generally refer to a specific commodity or service in any of these three ways. Although a specific money object can contextually evoke, for instance, the labor or sale that it represents, or its donor, or the monarch or nation whose imprint it bears, or even the purchase it is destined to perform, its fundamental property is its capacity to assume any meaning at all that its owner bestows upon it. This is tantamount to saying that money is a sign without meaning, that is, without a referent (cf. Rotman 1987). This semiotic property of money is undoubtedly the feature that qualifies it as both the most celebrated and the most condemned of human inventions.

A second, related observation is that the code by which money communicates information only has one character. This is concomitant to the observation that the money sign can stand for anything at all, which means that there is nothing that it can be opposed to. Other kinds of codes (such as alphabets, genetic codes, musical scores) have more than one character, which is a basic requirement for transmitting information. It could be objected that the *absence* of money constitutes a binary opposite to its presence, so that a money payment can be interpreted as a message encouraging whatever activity is being paid for, while its absence would discourage it, but the undifferentiated character of money cannot convey messages more meaningful than a signal to continue whatever is being done. It can be argued that this limitation has important implications for sustainability. In principle, the parallel existence of two distinct currencies pertaining to separate kinds of exchanges would grant market actors the capacity to transmit messages about the limits of commensurability and thus about the range of possible repercussions that may result from their transactions.

A third observation is that even if money is conceded to signify nothing but abstract quantity, such signification will mean very different things to different people, depending on the amount of money they have at their disposal. This inherently asymmetrical aspect of commercial transactions completely contradicts the liberal image of the generalized and unregulated market as free, fair, and of universal benefit (Reddy 1987: 62–106). Asymmetrical exchange is certainly not specific to money-based economies, but money is a way of concealing such asymmetries by couching them in an idiom projecting the appearance of reciprocity and fairness. This intrinsic asymmetry between market actors, inherent in their divergent assets, applies regardless of whether there are asymmetries in the physical substance of the exchange.

A fourth observation on the peculiarity of money is that "it is a form of social power that has no inherent limit" (Harvey 2010: 43). There is always a limit to the amount of physical assets a person can own, but there is no inherent limit to the amount of money he or she can command. Thus, there is no limit to the amount of money a human can desire. This is another way of phrasing the implications of the mainstream abandonment, within economics, of concerns with the finite, material aspects of the economy. As conceptualized by neoclassical economics, the economy can "expand without getting physically bigger" (Mitchell 2009: 417). The gross national product was invented to measure "the speed and frequency with which paper money changed hands," and it "could grow without any problem of physical or territorial limits" (ibid.: 418).

Money, Value, and the Material: From Political Economy to Economics

There is a wide consensus that modern economics has emerged as the understanding and explanation of capitalism (Heilbroner 1999 [1953]). Although money, market exchange, and price relations have existed for millennia, it was the conceptualization of abstract land, labor, and capital as quantifiable and commensurable categories that created the discipline of economics (ibid.: 27). The emergence of economics has thus reflected and reinforced historical processes of commercialization and monetization. Although various schools of economics advocate different economic policies, they share the underlying assumption that (general-purpose) money is a valid metric for quantifying human transactions, and that statistics and mathematics offer methods for thinking and deliberating about them (ibid.: 314).[2] Significantly, a standard textbook on the history of economics that has shaped the minds of generations of economists does not devote a single word to reflecting on the phenomenon of money itself, without which economics as a discipline would not exist.[3]

The expansion of market trade in the late Middle Ages and early modern period contributed to a long-standing confusion about the relation between what Aristotle had called "use value" and "exchange value." The intuitive distinction between a traded commodity's substance or particular qualities and its abstract and quantifiable exchange value is valid, but rather than acknowledge that these two features of a commodity are analytically incommensurable, the history of economic thought has been plagued by unsuccessful attempts to derive the latter from the former. The challenge, as visualized by economists from the Physiocrats through Marx to ecological economics, has been to relate the material aspects of economic processes to the accumulation of monetary value. In the preindustrial world of the Physiocrats, it was the fecundity of the soil that was the ultimate source of wealth. For the economists of the Industrial Revolution, such as Ricardo and Marx, it was the power of labor. For many modern ecological economists, it is energy and "natural capital" (cf. Martinez-Alier 1987; Costanza et al. 1997). What these perspectives have in common is the notion that some particular, physical input in the production of a commodity has a specifiable relation to the monetary income from selling it. The urge to relate money and accumulation to tangible, biophysical realities is commendable—and an expression of a widespread dismay at seeing them diverge in both thought and practice[4]—but any attempt to derive the former in determined and definable ways from the latter is misguided. There is an important difference between observing that economic processes that augment exchange values and monetary accumulation simultaneously imply the dissipation and degradation of natural resources, as was demonstrated by Nicholas Georgescu-Roegen (1971),[5] on the one hand, and proposing theories of economic value based on land, labor, or energy, on the other. The real challenge for an economics concerned with sustainability ought to be how to respond to the problems posed by Georgescu-Roegen's observations, while avoiding the pitfall of advocating a materialist theory of value.

In view of the problems confronted by attempts to derive the economic from the biophysical, it is understandable that mainstream economics as a discipline and a profession has more or less abandoned such attempts. The logic of market exchange explored, for instance, by Alfred Marshall—whether identified in the exchange rates negotiated by Andean peasants at a rural vegetable market or in those used by Wall Street stock brokers—is an issue quite distinct from questions of the biophysical substance of commodity flows, material asymmetries in trade, accumulation, and inequalities. The negotiations of exchange values are conditioned by contextual factors—such as supply, demand, and relative purchasing power—that are distinct from the material content of exchange. The former—the intricate mathematics of market equilibrium—is what mainstream economics is all about (Heilbroner 1995 [1953]: 210). The latter are concerns of classical political economy, raised today only by heterodox schools such as Marxism and ecological economics.

A brief review of some of the main contributions to modern economic thought will illustrate this abandonment of materiality in mainstream economics. Whereas François Quesnay and the Physiocrats argued that only agricultural labor, assisted by nature, produced economic value, Adam Smith recognized labor in general as a source of wealth. Both positions were related to morality in the sense that they built on notions of productive versus unproductive activities. Smith's most fundamental argument, however, was that the market exchange of commodities promoted a fair and efficient allocation of goods at optimal exchange rates that were of benefit to all participants. Ricardo extended Smith's understanding of the virtues of the self-regulating market to the international arena, and elaborated his shift to a labor theory of value. Both Smith and Ricardo attempted to show how market prices (exchange values) reflected the amount of labor time that had been invested in a commodity. In encouraging the displacement of British land requirements overseas, for instance, through his criticism of the protectionist Corn Laws, Ricardo further contributed to the conceptual developments accompanying the transition from agricultural to industrial society. This was importantly reflected in his proposition that the factors of production (land, labor, and capital) are substitutable for each other, an assumption that has been rejected as fundamentally flawed by the field of ecological economics (cf. Daly 1996). In effect, mainstream economics following Ricardo has assumed that the Industrial Revolution, by displacing resource requirements overseas as well as underground, had dissolved the age-old land constraint (Wrigley 1962; Wilkinson 1973; Mayumi 1991; Pomeranz 2000; Sieferle 2001 [1982]).[6] Also reflecting actual social processes in his time, Ricardo envisaged technological improvements as a source of economic expansion far beyond the constraints imagined by Smith. The ideology of "cornucopianism" that pervades modern economics can be traced to "the bonanza of New World settlement and fossil fuel energy" (Albritton Jonsson 2014: 168).

Marx's understanding of technological progress and industrialization also built on a labor theory of value and the industrialists' desire to increase their profits by lowering their wage costs. His assertion that industrialists derive their profits from the exploitation of labor can be traced to ideas of earlier economists, such as the urge of the Physiocrats to identify a factor of production—in their case, land rather than labor—that generates more economic value than it costs, and the awareness of Ricardo, Thomas Malthus, and John Stuart Mill that the rate of capital accumulation is constrained by wage levels.

Although Marxian and (neo-Physiocrat[7]) ecological economics have continued to challenge mainstream or neoclassical economics (represented by, for instance, Marshall and León Walras), the general economic expansion in Europe and North America since the late nineteenth century has encouraged the economic profession since the so-called marginalist revolution to largely confine itself to mathematical calculations of market equilibrium, rather than focus on the causes of inequality. To be sure, prominent economists during the past century have addressed a number of highly controversial issues—from John Hobson's critical understanding of imperialism to John Maynard Keynes's advocacy of government intervention—but the profession has been increasingly united by shared assumptions about the long-term benefits of general-purpose money, globalizing markets, and technological innovation, and about the usefulness of abstract thought experiments, diagrams, and algebra in representing economic behavior. Most significantly, in this context, very few mainstream economists today would concern themselves with any other metric than the money gained through market exchange. The sophisticated mathematical methods of economics are applied to this metric only, to the virtually complete exclusion of material metrics such as hectares of land, hours of labor, or Joules of energy. Although flows of biophysical resources such as embodied land, labor, and energy are unquestionably relevant to issues pertaining to sustainability, they have been expelled from the perspective of neoclassical economics. For instance, whereas William Stanley Jevons in 1865 had expressed concern regarding the future exhaustion of coal reserves, Keynes in 1936 believed that they could be replaced with reserves of currency (Mitchell 2009: 416).

Money and Morality: Historical and Cross-Cultural Perspectives

The emergence of general-purpose money has been recursively connected to the emergence of modern forms of social life and thought (Simmel 1990 [1907]). Through centuries of discussions about the social implications of these processes, a central theme has been the relationship between money and morality. Already in the fourth century BC Aristotle denounced moneymaking for its own sake (i.e., *chrematistics*), and four centuries later St. Paul warned that "the love of money is the root of all evil," but the sin of avarice seems to have been particularly condemned from the expansion of market trade in the eleventh to thirteenth centuries (Bloch and Parry 1989: 18). Aristotle's position was revived in the thirteenth century by Thomas Aquinas, who classified avarice as a cardinal sin, and up until the eighteenth century, the official condemnation of moneymaking in European civilization ran parallel to its increasing centrality in economic life (Macfarlane 1985: 69–71). Central to these deliberations were attitudes toward the phenomenon of interest. As is reflected in several of Shakespeare's works, money blurs the moral distinction between good and evil (ibid.). From the late Middle Ages, avarice was viewed as less and less sinful (Hirschman 1977), and in 1714 Bernard Mandeville's *Fable of the Bees* finally equated "private vice" with "public benefit," which ever since Adam Smith's *The Wealth of Nations* has been the fundamental creed of economics (cf. Dumont 1977: 63). The five centuries between Aquinas and Smith saw an unprecedented expansion of commerce and ultimately the promotion of moneymaking from vice to virtue. As Maurice Bloch and Jonathan Parry (1989: 29) argue, in capitalist ideology, "the values of the short-term order have become elaborated into a theory of long-term reproduction." Another way of putting this is that "economics had to emancipate itself from morality" (Dumont 1977: 36). Economics has detached itself from ethical considerations, even though this has often entailed a distortion of Adam Smith's own views on ethics (Sen 1987).

As David Graeber (2011a) has shown, however, economic obligations generate their own varieties of rationality that are, at different levels, simultaneously framed by and separated from issues of morality. The historical inclusion of human obligations in the sphere of goods exchanged through the medium of general-purpose money has generated pervasive ambiguities about how to draw boundaries between persons and commoditized things, as drastically illustrated by the phenomenon of slavery.[8] Drawing on several millennia of human history, Graeber shows that societies in which economic indebtedness grows to the point where it more or less literally enslaves major parts of the population tend to reach thresholds where morality again intervenes in economics and there are large-scale cancellations of debt. In the normal operation of such economies, however, the mechanical rationality of managing money tends to be decoupled not only from considerations of face-to-face human morality, but from the exigencies of living sustainably on planet Earth. Not least in the Marxian tradition, the logic of capitalism is recognized as inherently opposed to sustainability (Foster et al. 2010; Klein 2014).

It is important to consider the connection between the two kinds of detachment that mainstream economics has achieved over the past two centuries, and that we have addressed above: the detachment from material processes and from morality. As Aquinas's condemnation of moneymaking was based on his conviction that merchants and moneylenders do not create value as laborers do, there is an interesting line of descent from Aquinas to the labor theory of value (Bloch and Parry 1989: 3, reference to R. H. Tawney). It is thus no coincidence that schools of economics which today have moral objections to what they identify as forms of unequal exchange that are invisible to mainstream economists—primarily Marxian and ecological economics—are precisely those schools that maintain a strong concern with material processes. It appears that arguments appealing to moral norms such as justice and equality need to be based on real asymmetries in the flows of embodied biophysical resources, whether labor time, hectares of land, or Joules of energy. It seems very significant that neoclassical economics is as impervious to moral arguments as it is to material metrics.

Robert Heilbroner (1995 [1953]: 317–318) concludes the final edition of his classic textbook by showing how unreasonable it is of economists to claim to conduct scientifically objective studies of human volition, the distribution of wealth and income, and other "highly mutable determinations of the sociopolitical order in which we live." What does it mean, he asks, to be "objective" about "such things as inherited wealth or immiserating poverty?" Although he does not include general-purpose money and globalized market exchange in his category of mutable sociopolitical arrangements, this is an inescapable implication of an anthropological perspective.

Though an economist rather than an anthropologist, Thorstein Veblen managed in 1899 to defamiliarize everyday behavior in American society by exposing its cultural and sociological aspects. Veblen provided a theoretical foundation for the school of institutional economics, which continues to challenge the assumptions and scientific aspirations of neoclassical economics. He pioneered cultural and sociological analyses of consumption such as those much later presented by Jean Baudrillard (1981 [1972]) and Pierre Bourdieu (1984 [1979]). In the economic anthropology of the 1950s and 1960s, the debate between institutional and neoclassical economics was duplicated by the debate between substantivists and formalists, in which the former tended to emphasize the cultural specificity and contextuality of economic behavior, while the latter claimed to be able to identify similar patterns of calculation and maximization regardless of cultural context.[9]

In 1944, Karl Polanyi (1957 [1944]) published his influential book *The Great Transformation*, which critically traces the social ramifications of the emergence of market capitalism in nineteenth-century Europe. Polanyi demonstrates how the establishment of economics as a discipline and worldview can be understood as a cultural process accompanying and reinforcing historical transformations in economic practices. The ambition to understand modern economic thought as a cultural system, that is, to approach the discipline of economics using the conceptual tools of anthropology, also characterizes Marshall Sahlins's *Culture and Practical Reason* (1976), Louis Dumont's *From Mandeville to Marx* (1977), and Stephen Gudeman's *Economics as Culture* (1986). Moreover, several anthropologists have dealt with the general topic of money as a cultural phenomenon (for instance, Crump 1981; Parry and Bloch 1989; Weatherford 1997; Hart 2000; Graeber 2011a).

Heilbroner's (1995 [1953]: 308–310) point, inspired by Joseph Schumpeter, that individual economists see things the way in which they wish to see them, is perfectly illustrated by the diametrically opposite views of capitalism presented by Marx and Polanyi, on the one hand, and by mainstream economists (including Heilbroner himself), on the other. For Polanyi, as for Marx, the emergence of the disembedded market economy is a tragic story of human suffering, while for Heilbroner and most of his colleagues it is a story of emancipation: the very commoditization and abstraction of land, labor, and capital that Polanyi laments is for Heilbroner what liberates economic logic from the fetters of social bonds, politics, religion, and culture (ibid.: 24–29).

Polanyi and George Dalton were leading proponents of the substantivist school in economic anthropology, as was Paul Bohannan (1955), who in the early 1950s identified the existence of separate spheres of exchange and special-purpose currencies among the Tiv of northern Nigeria. Although the ethnographic facts and explanation of such "multicentric" economies have been disputed, the very idea of distinguishing between separate spheres of value is worthy of reflection and consideration (see chapter 8). The fundamental problems of global sustainability may not be inherent in the market principle in itself as much as in the implications of general-purpose money and the globalized *scale* of the market. A way of curbing the destructive consequences of economic globalization might be to rediscover the virtues of distinguishing local values (such as those concerned with food, shelter, energy, place, community, and face-to-face relations) from the values pertaining to global communication. Suffice it to say, at this point, that these virtues would be very difficult to grasp from the perspective of mainstream economics.

Financial Crises from Precious Metals to Virtual Money

It is no coincidence that Aristotle's moral objections to moneymaking appeared in the first truly commercial civilization, established in the Aegean area several centuries BC (cf. Weatherford 1997: 28–45). The metal coinage that was invented in the region around 700 BC undermined the ancient agrarian, tributary empires and provided the foundations of the so-called Axial Age (700 BC–AD 600). The transition from credit money, built on trust, to commodity money (precious metals) encouraged warfare, plunder, and slavery in this period (Graeber 2011a). The Middle Ages (AD 600–1450) saw a return to credit money and tribute in kind accompanied by a cosmological emphasis on material production, rather than money itself, as the source of value, but the introduction of paper notes in Renaissance Italy in the fourteenth century initiated the transition from feudalism to modern forms of banking and capitalism. From the late fifteenth century, the early modern capitalist empires again focused on precious metals, epitomized by the doctrine of mercantilism. The worldview of the eighteenth-century Physiocrats retained a feudal emphasis on the material fecundity of land, but adopted an abstract analytical framework for understanding economic processes that was later to be conducive to conceptualizing the productivity of labor in early industrial Britain. As already noted, the labor theory of value thus traces its roots to medieval church doctrine and ultimately Aristotle, as opposed to the age-old inclination toward money fetishism, which has been particularly pronounced in periods emphasizing commodity money, such as the Axial Age as well as the period of capitalist empires after 1450 (ibid.).

The year 1971 marks the advent of electronic money and an electronic stock market (National Association of Securities Dealers Automated Quotations [NASDAQ]) as well as the abandonment of the Bretton Woods gold standard. Since then, there has certainly been a resurgence of credit money, as Graeber observes, but rather than an emphasis on material production (as in the Middle Ages) we have witnessed a further emancipation and fetishization of autonomous monetary value. It remains to be seen whether the events of 1971 were really another turning point in the grand historical oscillations identified by Graeber, or a more temporary incident. The general historical trend toward a transition from metal through paper to electronic money has entailed a progressive separation of finance and monetary flows from real flows of matter and energy. Recurrent attempts to discipline banks and politicians, constraining them from issuing excessive amounts of new money by tying major currencies to a metal standard (for instance, by the Bank of England in 1844, the US Congress in 1900, and the agreement at Bretton Woods in 1944/1946) have all ended in a similar way. As the amount of paper currency in circulation has increased, diverging more and more from the value of a finite stock of bullion, the end result has repeatedly been devaluation and the severance of metal standards.[10] The volatility of trust as the sole foundation of economic value has led to recurrent financial breakdowns, from the banks of Florence in 1343 to the Wall Street stock markets in 1929 and 2008.

The final severance of the connection between the leading global currency and the physical substance of gold in 1971[11] once again signaled that more money was being issued than could be backed by bullion. As had occurred repeatedly in history, the widened scope for credit money opened a new era of financialization, in which the incentive to increase profits on capital was increasingly disconnected from production processes involving the physical use of labor and land. Capital itself became the most profitable of production factors, in Marxian terms signifying a shift from M-C-M¹ to M-M¹. From a semiotic perspective, financialization implies that money increasingly refers only to itself, rather than to some material substance or activity. The preoccupation with the apparent capacity of money to grow on its own is tantamount to fetishism, and raises the same intuitive concerns about the unproductiveness of chrematistics as were voiced already by Aristotle. What both the Third World debt crisis in the 1980s and the Wall Street financial crisis in 2008 have made abundantly clear is that the vicissitudes of finance capital can have extremely serious consequences for land and labor. The so-called structural adjustment policies of the 1980s coerced less affluent countries to intensify the pressure on both land and labor (Körner et al. 1986; Altvater 1990: 28–31; Stevis and Assetto 2001; Ravenhill 2005), and the austerity measures prompted by the financial crises in the United States and Europe since 2008 have been devastating for significant parts of the population even in more affluent nations. These crises have renewed the concern with how human welfare and environmental health might be insulated from the unpredictable cybernetics of finance capital, as reflected in several initiatives to promote complementary currencies. They have also renewed concerns about the extent to which money should be permitted to disconnect from physical reality—whether a stock of gold or some other finite quantity—prompted by a widespread conviction that it is such disconnectedness that is the root cause of financial crisis. Several contributors to these debates have suggested that financial crises reflect structural problems of a tangible, material nature, such as declining returns on inputs in energy production (Hall and Klitgaard 2011; Lipson 2011; Nikiforuk 2012; Tainter and Patzek 2012).

Mainstream neoclassical economics do not search for material causes of financialization and financial crises. Neither the conclusions of the US Financial Crisis Inquiry Commission (2011), the letter from the British Academy to Queen Elizabeth II on the global financial crisis (Besley and Hennessy 2009), nor a "Nobel Prize" winner's analysis of it (Stiglitz 2010) identify the relation between money and material aspects as a relevant factor. These mainstream accounts emphasize the easy accessibility of inexpensive credit, insufficient societal regulation (based on an ideological faith in the self-regulating market), human weakness (wishful thinking, hubris, greed, fraud, and irresponsibility), and the lack of comprehensive understanding of the financial system as a whole. If we disregard the role of human weakness, which is no doubt an invariant and incorrigible factor contingent on societal frameworks for action, the mainstream perception of the problem appears to boil down to insufficient regulation of credit and financial risk management, but in acknowledging the element of surprise (cf. Queen Elizabeth's pertinent question, "Why didn't anybody notice?"; Besley and Hennessy 2009) there is also a concession that mainstream understandings of the economy were deficient. For Joseph Stiglitz (2010), at the policy level this is basically the old disagreement between advocates of Keynesian government intervention (to which he belongs) and free-market fundamentalists regarding the best way of increasing aggregate demand.[12] At the level of transactions, the new financial instruments introduced in the years leading up to the Wall Street crisis were a means to spread the risks and to "exploit the poor," but they simultaneously created "new problems of imperfect information" (ibid.: 14–15).[13]

Indeed, the conclusions of the Financial Crisis Inquiry Commission itself suggest that, even to this presumable authority on economic matters, the logic of the crisis remains largely opaque. Its 14 pages are saturated with diffuse metaphors that would hardly be acceptable in other fields of social science. In these few pages aiming to summarize the crisis, we are told that there were "toxic" mortgages, "red flags," and a "highway [with] neither speed limits nor neatly painted lines," while financial speculators were "flying ever closer to the sun." Then the "bubble burst" when the "spark that ignited" events "lit and spread the flame" that caused the system to "race ahead" of our ability as we "reaped what we had sown." There was a "big miss, not a stumble" and although economists tried to "put fingers in the dike" and build a "bulwark" against panics, "the contagion spread" and the "rush for fool's gold" simply "added helium to the housing balloon" (Financial Crisis Inquiry Commission 2011: xv–xxviii). It is doubtful whether these metaphors are helpful in clarifying, to noneconomists, the causes of the crisis and the economic theories that are believed to account for it. Such language represents an inside view of the American economy in which the rules of the monetary game are taken for granted rather than analyzed, for instance, in relation to global processes or to material aspects.

Even in Stiglitz's (2010) much more informative and globally oriented account, comparatively little space is devoted to perspectives that transcend the deliberations on how money should be managed on Wall Street. He does mention, however, among factors that have contributed to the crisis, the rise in oil prices between 2003 and 2008, growing global inequalities that "shifted money from those [who] would have spent it to those who didn't," the policy in less developed and oil-producing countries to accumulate reserves, and the increasing economic globalization which made it possible for the United States to so excessively live beyond its means, as "the world's consumer of last resort" (ibid.: 4, 19, 20, 25). Several economists have gone beyond the Keynesian critique of what Stiglitz calls free-market fundamentalism to argue that the stock market crash of 2008 was also the crash of neoclassical economic theory (Keen 2011; Mirowski 2013).

Marxian perspectives on financialization and financial crises tend to emphasize the structural contradictions and trajectories of capitalist systems of economic accumulation (Foster and Magdoff 2009; Harvey 2010; Screpanti 2014). Critically reconsidering the distinction between finance and fictitious capital, on the one hand, and the real economy of production,[14] on the other, Marxian economists have addressed the complex relation between money as sign and the material substrate of goods and services to which it ideally might be presumed to refer. Finding it difficult to clearly distinguish the two, John Bellamy Foster and Fred Magdoff (2009: 7) conclude that "both production and finance under capitalism are at one and the same time both real and monetary in nature," but they follow Paul Sweezy in observing that there has been an inversion between the financial and the real (ibid.: 81). Financialization is explained in terms of the "search by capital for profitable outlets for its surplus despite the stagnation of investment opportunities within production" (ibid.: 18).[15] The stagnation of production is in turn explained by the problem of overproduction (sometimes referred to as underconsumption), generated by the contradictory imperative of capital to keep "wages down while ultimately relying on wage-based consumption" (ibid.: 27). Although experiencing a decline in real income from 2000 to 2004, American households were able to increase their consumption by utilizing easy credit and thus increasing their debt (Harvey 2010: 17). Their increasing access to credit was made possible by a combination of excess capital (offering minimized interest rates and other service costs on loans), high real estate values (facilitating mortgage), and creative "securitization" by the financial institutions (repackaging and transfer of risk in the form of increasingly opaque "financial instruments," reducing creditors' worries about default). Gains from such financial activity began to exceed profits from manufacturing already in the early 1990s. Capital thus used the vast surplus at its disposal not to invest in new productive capacity, but to increase its financial claims to wealth (Foster and Magdoff 2009: 60). From a Marxian perspective, "the huge expansion of debt and speculation provide ways to extract more surplus from the general population and are, thus, part of capital's exploitation of workers and the lower middle class" (ibid.: 61). Marxian analysts tend to view the financial crisis that struck the United States in 2008 as a symptom of declining global hegemony (Foster and Magdoff 2009: 22, 75; Friedman and

Ekholm Friedman 2013; Kalb 2013). There is also an awareness of the significance of US oil imports in the nation's escalating foreign trade deficit and debt (Harvey 2010: 79).

If, as is often suggested, the current era of financialization was inaugurated by the abandonment of the Bretton Woods agreement in 1971, it is interesting to consider the proposal that the real physical standard on which the value of the dollar had been based prior to that date was actually not gold, but oil (Mitchell 2009: 414, 2011). A more general variety of this argument, presented by several proponents of ecological economics, is that the financial crisis is a reflection of the disjunction between the fictitious paper economy and the genuinely real economy of energy and material flows (Kallis et al. 2009; Hall and Klitgaard 2011; Heinberg 2011; Lipson 2011; Daly 2012; Townsend 2013; Smith-Nonini 2014). This approach, which can be traced at least back to Frederick Soddy (1926), observes that "underneath the economists' real economy, there is the ecological economists' *real-real* economy, the flows of energy and materials whose growth depends partly on economic factors (types of markets, prices) and in part from [*sic*] physical and biological limits" (Kallis et al. 2009: 16).

Mainstream and heterodox economic theories appear to agree about some of the problems exposed by the financial crisis: in particular, the excess of capital in relation to demand, the absence of regulation of credit, the opaqueness of financial instruments for transferring risk, and the large extent to which financial policy derives from the economic interests of the financial sector rather than from those of the majority of people, or from rigorous analysis. Several analysts of the crisis claim to have predicted it, but from very different vantage points (Foster and Magdoff 2009; Stiglitz 2010; Keen 2011). Much of the literature is couched in language referring to the chrematistic technicalities which preoccupy the financial institutions on Wall Street, that is, the opaque "financial instruments" which were *designed precisely to confuse*. It is thus not surprising that few external analysts are able to penetrate the logic of these complex modes of transferring economic risk. Their operation tends to be as inaccessible to outsiders as the details of advanced engineering, and as conducive to asymmetries of social power. In several respects, the language of Wall Street—and of the discipline of economics as a whole—represents a privileged domain of discourse that suggests intentional obscurantism, much like the intricacies of medieval theology. This lack of transparency is fundamentally at odds with ideals of democracy, accountability, and equality. The mysteries of recent financial speculation have revealed the ultimate implications of the magic of money—as a social game in which only a small minority know the rules, and the great majority are losers.

CHAPTER 4

Empires, World-Systems, and Expanding Markets

In 1526, Francisco Pizarro's pilot Bartolomé Ruiz was sailing south along the coast of Ecuador when he encountered a native raft from Peru loaded with the reddish shells of the Thorny Oyster, or *Spondylus*. The Spanish conquistadors were looking for gold and silver, and could not understand why the natives would make long trading voyages in pursuit of seashells. It was an encounter not only between two vessels, but between two empires with different cultural notions of exchange value, geared to different conceptions of the magical agency of particular artifacts and substances. The Spaniards were heading south in search of yellow and gray metals, while the native Peruvians were heading north in search of red seashells. They had more in common than either of them could imagine.

The Spanish Empire was largely based on the extraction of precious metals from the New World. These metals were the very substance of money in the Old World, and silver, in particular, was in very high demand because it was the primary means for Europe to conduct trade with distant China. The import of prestigious silk and porcelain from China to Europe had to be paid for in silver, because this was what the Chinese demanded in exchange. Ultimately, it was the demand for silver in China that propelled the Spanish conquests in Mexico and Peru.

The Inca Empire did not use money in the European sense, but gifts and sacrifices of prestigious valuables were central to its organization. Like silver in the Old World, *Spondylus* was a rare and fetishized substance that was used to regulate the exchange of goods and services in accordance with cultural conceptions of reciprocity. Even the sacrifices of *Spondylus* to Andean gods were part of the flows of gifts which constituted the Inca economy.

In this chapter, we examine the different ways in which hierarchical societies have appropriated resources from the geographical areas into which they have expanded, and how they have employed specific kinds of artifacts to achieve such appropriation. The exploited resources include both human labor time and the products of the land. Although the concept of imperialism is often reserved for territorial expansion and political control, a central point of this chapter is that markets can provide means of appropriation which are just as compelling—although less apparent and less demanding in terms of coercive infrastructure—as the political incorporation of exploited territories.

The Ecology of Empire

If an empire is defined as an expansive state striving to extend its geographical territory by engulfing other states or nonstate societies, an underlying objective is obviously to gain control over the land and labor assets of such territory. Land and labor are convenient concepts for productive resources deriving from natural space and investments of human labor time, respectively, but their analytical separation should not obscure their interdependency. Much of what we think of as land (for instance, agricultural areas, managed forests, mines) represent considerable past investments of human labor, whereas labor is inconceivable without the food energy and other resources drawn from land. Both these factors of production, in other words, are inextricably bound to ecology.

In examining actual processes of imperial expansion, it is thus possible to compare different ways in which states have appropriated ecological resources from their hinterlands. Such a comparative framework can deal with qualitative differences such as institutions, ideologies, technologies, or consumption patterns, or it can attempt to quantify the flows of embodied (or invested) land and labor measured in hectares and hours of human time, respectively. The latter approach is particularly useful if the aim is to understand imperial systems in terms of their material metabolism, as measures of the productivity of land and labor can be converted into measures of energy (Smil 1994). However, it should rarely be necessary to calculate the quantitative details of imperial energy flows in order to offer analytical observations on the metabolic organization of empires founded, for instance, on the appropriation of human and animal work, food, and fodder.

In trying to understand societal structures such as empires in terms of natural aspects such as biogeography or energy, we need to address the relation between society and nature. This chapter will argue for an understanding of socioecological systems that acknowledges the significance of perspectives from both social and natural science, but that is strongly critical of biogeographical determinism. The risk of introducing tangible, physical parameters such as energy or hectares of agricultural land is that it may inspire some readers to think of such aspects as causally primary in a simplistic sense, as it may give the impression of denying the complexity of cultural specificities and historical contingencies in order to reveal their ultimate evolutionary "function" (cf. White 1959). The challenge, in other words, is to understand the material dimension of imperialism without reducing sociocultural projects to reflections of material exigencies.

A further challenge, also addressed in this chapter, is to understand different varieties of imperial control and appropriation of material resources as embedded in cultural ideologies that represent unequal exchange as reciprocal, or at least fair (Godelier 1986). This approach should help us grasp the ideological role of mainstream modern conceptions of national economic growth and technological progress. It will simultaneously give us reason to reflect over the conspicuous absence, in the voluminous literature on the political and economic history of imperialism, of critical scrutiny of the global, social, and ecological implications of these taken-for-granted notions. In particular, this chapter emphasizes that the ecological implications of imperialism are much more politically and morally charged than the (largely unintentional) diffusion of plants and animals.

"Ecological Imperialism" Revisited

In his 1986 classic *Ecological Imperialism: The Biological Expansion of Europe, 900–1900*, the historian Alfred Crosby argued that the success of European imperialism in temperate areas of the world was largely a consequence of the successful expansion of a biological assemblage of species that accompanied the European colonizers, displacing indigenous peoples, plants, and animals. Although the crucial role of epidemics in European expansion has long been recognized, Crosby's main contribution was to show how the microbes were part of a "portmanteau biota" that existed in symbiosis with Europeans and that also included domesticated crops and animals as well as weeds and pests. The biogeographical success of this biota, including European humans, is presented in Darwinian terms, as a process of selection ultimately determined by natural phenomena such as primeval tectonic shifts and the long-term isolation of biological populations. About 3,000 years ago, Crosby (1986: 34) suggests, "the human of Old World civilization" was something of a "superman," in that he served as the "template" for all humans destined for global expansion. The features that would prove to be of such advantage to these Old World humans included the ability not only to cultivate food and fiber, domesticate animals, and use the wheel, but also to coexist with weeds, vermin, parasites, and a variety of disease-carrying microbes.

A similar argument was elaborated 11 years later by the geographer Jared Diamond in his 1997 bestseller *Guns, Germs, and Steel: The Fates of Human*

Societies. Like Crosby, Diamond traces the ultimate roots of European dominance in the modern world to biogeographical circumstances several millennia back in time. In attempting to explain the unequal global distribution of wealth and power, Diamond is careful to reject explanations referring to biological differences between human populations, but his account nevertheless refers to natural factors that in the final instance render social-science approaches superfluous. Although he tries to evade charges of Eurocentrism and the justification of domination as inevitable, Diamond's account—like Crosby's—tends to naturalize European expansion as the outcome of physical conditions and inexorable Darwinian processes of selection. These physical conditions include the geographical orientations of continents and the distribution of various species of wild animals and plants. Like most other evolutionists (cf. White 1959; Morris 2010), Crosby and Diamond illustrate how an interest in the material aspects of historical processes tends to go hand in hand with Darwinian narratives that naturalize power.

Neither Crosby nor Diamond offers any significant insights into the role of macrosociological processes in the economic history of European expansion. In fact, it seems to be precisely by avoiding social-science theorizing in favor of rather simple and easily grasped biogeographical models that they have gained such wide readerships. But social science does offer more credible accounts of societal expansions than biogeography. As unthinkable as it would be for scholars of economic history to disregard physical factors such as transportation routes, energy sources, demography, or farmland, it is futile for biogeographers to try to account for European expansion without considering, for instance, the cultural patterns of consumption encouraging specific constellations of long-distance trade. Such consumption patterns are idiosyncratic, historical accidents generated by more or less arbitrary semiotic systems that assign specific social significance to particular trade goods such as certain kinds of food, textiles, porcelain, or metals (Sahlins 1976; Bourdieu 1984 [1979]; Baudrillard 1981 [1972]). Empires do not simply pursue foodstuffs defined by human metabolism, but very specific symbolic values such as the *Spondylus* shells coveted by the Inca or the myrrh retrieved from distant Punt by Queen Hatshepsut of Egypt.

The economic and political expansion of Europe was thus indissolubly linked to its demand, for instance, for silk, spices, beaver pelts, and sugar, and to the Chinese demand for silver. Andre Gunder Frank (1998) and Kenneth Pomeranz (2000) persuasively argue that, rather than being predetermined by geographical circumstances several millennia ago, the rise of the West was an accident of late eighteenth-century history. The European peninsula is a corner of the Old World that is close to the New World, and its history of expansion began when the two hemispheres were economically connected in the sixteenth century. Europe would, in time, draw great advantages from its proximity to the vast silver resources and conveniently depopulated lands of the Americas, but until the late eighteenth century remained inconspicuous in comparison to similar, densely populated areas of the Old World such as China and India. To ignore the role of such cultural and historical contingencies in accounting for European expansion must be regarded as a major omission. The critical question we should thus put to Crosby and Diamond is if the biological expansion of Europe really was a prerequisite to its economic expansion, or vice versa?

The approach adopted in this chapter is that, although biological factors such as epidemics certainly were significant in facilitating Europe's economic expansion, sociocultural incentives were the primary driving force. This is not to say that European society or culture was uniquely predisposed to expansion (cf. Blaut 1993, 2000), but that global conjunctures in the centuries following 1492 shaped the specific trajectory of Western imperialism. As this trajectory rests on modern conceptions of economic development and technological progress, the world to this day remains largely persuaded by the West's own narrative of its expansion. This means that to grasp the nature of imperialism in the modern world, it will be necessary to approach contemporary accounts of global inequalities as a specific kind of ideology that is comparable to the cultural constructions of earlier empires. Moreover, to recognize the ideological affinity of modern and premodern narratives of expansion, it will be necessary to unravel the various ways in which they justify the appropriation of material resources.

In order to understand imperialism as a recurrent socioecological phenomenon in world history, we need to ask how the expansion of European and modern empires after 1492 differ from as well as resemble earlier imperial expansions. Furthermore, we need to explore these differences and similarities in terms of both ideologies and material flows. By means of which narratives have empires justified their power, and in which ways have they appropriated the ecological resources of expanding territories? How, in other words, are cultural constructions and socioecological processes intertwined in world history?

As employed by Crosby, the concept of ecological imperialism has come to denote the ecological *consequences* of imperialism, rather than its ecological *rationale*. However, if we consider the asymmetric flows of energy within an empire, arguably as essential for its reproduction as its own narratives of expansion, it should be valid to think in terms of what I have elsewhere referred to as the "thermodynamics of imperialism" (Hornborg 1992). In an article titled *Ecological Imperialism: The Curse of Capitalism*, John Bellamy Foster and Brett Clark have explicitly proposed to replace Crosby's influential definition with a concern regarding "the growth of the center of the system at unsustainable rates, through the more thoroughgoing ecological degradation of the periphery" (Foster and Clark 2004: 198). Writing about the implications of peak oil for the position of the United States in the modern world, Foster (2008) has also used the concept of energy imperialism. We have every reason to think about imperial projects as ecological phenomena, involving increasingly long-distance appropriation of natural resources such as energy. In applying such a perspective to the modern world, however, we will inevitably be confronted with modern narratives proposing that imperialism is a thing of the past and a concept that is inapplicable to the current situation. This is a predictable aspect of the fact that we, like the subjects of empires in the past, tend to be impregnated with the specific imperial narratives of our time.

Toward a Comparative Study of Imperial Metabolism

The vast literature on the archaeology and history of empires contains a growing body of quantitative data on land use, tribute flows, and the organization of labor that could be used for a systematic, comparative analysis of the material metabolism of imperial projects. This chapter can merely suggest a possible direction for such work to pursue, building on some methodological experiments in quantifying asymmetric transfers of embodied labor and embodied land (Hornborg 2006; Warlenius 2011; Bogadóttir 2012). Labor and land can be translated as (human) time and (natural) space and their appropriation by imperial power centers understood as a societal redistribution of time and space. Further on in this chapter we see how modern technology fits into this picture, as a recent version of time–space appropriation, but the argument will make more sense if grounded in a long-term history of imperialism.

The list of imperial projects to be briefly considered here begins with two distinct and foundational Old World empires, Han China (206 BCE–220 CE) and Rome (241 BCE–476 CE). These imperial expansions are then compared with the two most prominent New World examples, the Inca (1438–1532) and the Aztec (1428–1519). We shall next consider the establishment of two transatlantic empires that were largely based on connecting Old and New World resource flows, the Spanish (1492–1780) and the British (1600–1947). Finally, we shall discuss continuities and discontinuities in imperial strategies through more than two millennia of world history, from ancient Rome to the United States. How can all these strategies be compared in terms of flows of appropriated energy, land, and labor, and how can they be differentiated in terms of ideologies and institutions?

Han China (206 BCE–220 CE)

After a civil war which ended the former Qin dynasty, the Han dynasty founded in 206 BCE consolidated imperial control over a Chinese population of around 60 million people and a territory extending about 4 million square kilometers (Scheidel 2009; Burbank and Cooper 2010). The metabolism of the empire was primarily based on the extraction of agricultural tribute, mostly in the form of grains such as rice, wheat, and millet. Based on an estimated average carrying capacity of two persons per hectare, the agricultural area can be roughly calculated at around 300,000 square kilometers. Oxen and water buffalo were used as draft animals, but all other work was carried out by humans. Along with grains, raw materials such as cotton and timber, as well as precious items such as jade, corals, ivory, pearls, and gems, were imported to imperial centers, which in turn exported and redistributed textiles (including silk), grains, incense, spices, medicines, and ornaments. The main forms of capital investments were hydraulic infrastructures for agriculture, transports (roads, canals), ceremonial architecture, and an army numbering several hundred thousand warriors. In ideological terms, imperial power was founded on the divinity of the royal dynasty and the perception of the empire as protection against enemies. In material terms, the fundamental rationale of the ancient Chinese Empire was to use military coercion and religious devotion to extract labor and food energy through tens of millions of peasants on hundreds of thousands of square kilometers of agricultural land.

Rome (241 BCE–476 CE)

By 241 BCE, the former city-state of Rome had begun to expand beyond the Italian peninsula by conquering the islands of Sicily, Sardinia, and Corsica (Scheidel 2009; Burbank and Cooper 2010; Woolf 2012). It then proceeded to conquer the Carthaginian territories in northern Africa and Iberia, consolidating an empire which controlled the Mediterranean and much of what is now Europe for several centuries, until the fall of Western Rome traditionally set to 476 CE. The total population of the empire was around 60–75 million people in 150 CE, of which around 1 million belonged to the city of Rome itself. Like Han China, it was basically an agrarian empire collecting tribute in grain, notably from Egypt. The 2.7 million hectares of agricultural land in Egypt supported an estimated 7.5 million people, suggesting a carrying capacity of 2.7 people per hectare. Unlike China, however, the Roman Empire depended heavily on maritime transports across the Mediterranean. Instead of the Chinese rivers and canals, Rome relied on the sea for bulk deliveries of grains to its center, perhaps as much as 400,000 tons in some years (Garnsey

60 • Global Magic

1988: 231–232; Debeir et al. 1991: 35). When imports were at their height, only around 10 percent of the wheat consumed in the city of Rome was grown in Italy (Myrdal 2012). Another difference was the central significance of slavery, an institution inherited from ancient Greece. To a large extent slaves were war captives from peripheral areas of the empire. It has been suggested that it was in part because they became increasingly expensive in the early centuries CE that some of their labor was replaced by water mills (Debeir et al. 1991: 39). Oxen were used as draft animals. Wine and olive oil were traded throughout the Mediterranean. Major capital investments were made in farmland, roads, architecture (including aqueducts), armies, and naval fleets. Although water mills and sailing ships represent alternative sources of energy, the Roman Empire was primarily based on the extraction of food energy from conquered territories and the exploitation of peasants and slave labor. Christianity became the state religion in the fourth century. In the year 395 the empire was split into a Western and an Eastern (Byzantine) half, the latter centered on the city of Constantinople. The Western half fell apart during the fifth century, but the Byzantine Empire survived for another millennium until Constantinople was conquered by the Ottomans in 1453.

Inca (1438–1532)

After a decisive victory over the neighboring Chanka polity in 1438, the kingdom of Cuzco in the southern Andes rapidly conquered the Andean highlands and Pacific coast of South America from northern Ecuador to central Chile (D'Altroy 2001, 2002). Military force remained a central means of expansion, but several provinces of the Inca Empire were integrated through diplomatic persuasion. At its height, prior to the Spanish conquest in 1532, the population of the empire may have been close to 12 million people. The majority were peasants and llama pastoralists, the main staples being potatoes (in the higher altitudes) and grains such as maize and quinoa. Estimates of the productivity of intensive raised-field potato cultivation in the Titicaca Basin suggest a local carrying capacity of up to four people per hectare. As there were no draft animals, all agricultural labor was carried out by humans. The Inca court established a redistributive system of tribute represented as relations of mutual benefit, in which common people delivered labor, military service, textiles, foodstuffs, and camelid fleece in exchange for ceremonially distributed gifts of maize beer, cloth, and ornaments. In addition to the maintenance of agricultural terraces, irrigation canals, armies, roads, and ceremonial architecture, the tribute was invested in a system of warehouses storing food and cloth for redistribution in times of need. The imperial ideology represented the Inca emperor as the divine son of the Sun and the source of agricultural productivity, which meant that the delivery of tribute to him was perceived as a compensation for harvests and for protection against famine. Investments in architecture and infrastructure built on the capacity of the imperial elite to extract labor, food, and materials from the territories over which it had gained control.

Aztec (1428–1519)

Following the defeat of neighboring Azcapotzalco in 1428, the Aztec polity centered on the city of Tenochtitlan had conquered much of what is now Mexico by the time of the Spanish conquest in 1519 (Smith 2001). About 250,000 hectares of raised fields and other intensively farmed land in the Basin of Mexico provided food for about one million people, suggesting a carrying capacity of around four people per hectare in the core area, a figure identical to the one for the Titicaca Basin. The main agricultural staple was maize. As in the Andes, there were no draft animals. In fact, in ancient Mexico there were not even any pack animals (like the Andean llama), which meant that all tribute and merchandise had to be carried by humans. The Mesoamerican area was considerably more commercialized than the Andes, and the political consolidation of Aztec hegemony was less formalized. Whether the extraction of goods from peripheral areas was perceived as tribute or trade seems unclear, as the emperor's tribute collectors were simultaneously merchants and much of the transfer of goods occurred at markets. Major items of trade and tribute included obsidian, cotton textiles, ceramics, feathers, gold, jade, turquoise, bronze, and cacao. Cacao beans and cotton textiles frequently served as currencies in market transactions. The Aztec Empire illustrates how the extraction of labor-power and natural resources for purposes of accumulation in core areas can occur largely through market (nonadministered) exchange, even in premodern contexts.

Spanish (1492–1780)

The Iberian peninsula was integrated into the Umayyad caliphate in 711 and remained partly Muslim territory until 1492, when the Habsburg dynasty defeated the last caliph of Granada and attempted (as Charlemagne had done seven centuries earlier) to resurrect the Western Roman Empire (Burbank and Cooper 2010; Altman 2011; Schwartz 2011). The ideology of conquest and unification was strongly grounded in Christian theology. Blocked by Muslim polities, which now controlled most of the Mediterranean, the Iberian rulers extended their interests westward, trading along the African coasts and establishing sugarcane plantations in Madeira, the Azores, the

Canary Islands, and in the Caribbean. Having sponsored exploration of the Caribbean since Christopher Columbus's first voyage in 1492, the Spanish Crown began extracting bullion from the Aztecs in the 1520s and from the Incas in the 1530s. By the end of the sixteenth century, Spanish silver mines in the New World yielded about 85 percent of the world's silver, but the Spanish Crown was heavily indebted, and much of the wealth ended up in the Netherlands, which was claiming independence from Habsburg Spain and Portugal at that time. Because of the strong demand for silver in China and India, the Dutch East India Company in the early seventeenth century was thus well equipped to assume control over the European trade in spices and other merchandise from Indonesia and the Indian Ocean. Previously controlled by the Portuguese, who had rounded Africa in 1497, this trade largely passed into Dutch hands over the course of the seventeenth century. Meanwhile, Portugal controlled much of the Atlantic trade in African slaves bound for its own sugar plantations in Brazil as well as for Spanish plantations in the rest of the Iberian New World. By the mid-eighteenth century, the Spanish colonial territories in the New World encompassed western South America (most of the continent except Portuguese Brazil), almost all of Mesoamerica and the Caribbean, and western North America. These territories exported several agricultural products such as sugar, tobacco, cacao, and hides, but precious metals remained most important. Spain itself provided its colonies with iron, mercury (for refining silver), wine, and olive oil, as well as luxury goods like textiles and porcelain that often had a non-Spanish origin. Spain also imported grains and timber from northern Europe. Its main capital investments were in ships, armies, and ceremonial architecture. The use of wind energy to propel its sailing vessels was a crucial complement to the muscle energy of humans, horses, and oxen in maintaining the metabolism of the empire. During the second half of the eighteenth century, the Spanish Empire experienced several major rebellions, such as the Túpac Amaru revolt aiming to resurrect Inca power in Peru in 1780.

British (1600–1947)

In the year 1600, the British East India Company was granted monopoly on English trade east of Africa. Although unable to compete with its Dutch counterpart in southeast Asia, it established successful relations with the Mughal Empire in India and soon exported Indian textiles, silk, indigo, saltpeter, and tea to Europe (Burbank and Cooper 2010; Chaplin 2011). Similar companies organized British expansion in North America and Africa: the Virginia Company (founded 1606) thus financed the establishment of English settlements in North America, beginning with Jamestown in 1607; the Royal Africa

Company (founded 1663) organized the trade in slaves for the British plantation colonies in the Caribbean; and the Hudson's Bay Company (founded 1670) controlled the North American fur trade. The island of Jamaica became an important British colony in 1655. In the seventeenth century, 60 percent of the people who crossed the Atlantic to live in the British, Spanish, Portuguese, French, and Dutch colonies in the New World were African slaves, a figure that rose to 75 percent in the next century. The slave trade and plantation system created demand for food exports from the British colonies in North America as well as for cotton textiles exported from Britain (Inikori 1989), while industrial workers in England derived a significant share of their food energy from Caribbean sugar mixed with tea from China and India (Mintz 1985). The imports to England of cotton, sugar, and other land-intensive products from the New World colonies were part of the so-called triangular trade through which British merchants exchanged cotton cloth and other manufactures for West African slaves, which were then shipped across the Atlantic and sold in the colonies. The development of steam technology originally prompted by the market demand for textiles simultaneously offered powerful new vehicles for transport and military conquest, most notably railways and steamboats. Although its colonial territories in southern North America (the future United States of America) declared independence from Britain in 1783, the empire continued to expand. In the early twentieth century, the British Empire encompassed around 650 million people and about 35 million square kilometers. Much of this territory remained part of the empire until the mid-twentieth century, when a number of important former colonies achieved independence, notably India in 1947 and several African nations in the 1960s. It was in the core of the British Empire that the Industrial Revolution occurred in the late eighteenth and early nineteenth centuries. This signified a new kind of capital investments alongside the fields, roads, canals, armies, ships, and architecture of earlier empires: capital accumulation in the British Empire also included factories, railroads, and steamships. The strategy of industrial imperialism established in Britain was to be adopted and expanded by its former colonies in North America, as the United States emerged as an imperial power in the mid-twentieth century. What these new investments in technology actually meant in terms of imperial metabolism is the topic of the next section.

Industrialization as a Strategy of Time–Space Appropriation

At the core of the Industrial Revolution was the substitution of human and animal muscle power, as well as water and wind power, with fossil energy, or—if you wish—the imperative to increase productivity per hour of human labor. Steam engines and steam-driven factories inaugurated this development toward what we now know as high-tech society and a continuous aspiration for economic growth. The most prominent economic thinkers from this period—including Thomas Malthus (1766–1834), David Ricardo (1772–1823), and Karl Marx (1818–1883)—all lived in England and devoted much effort to understanding the economic and technological changes of their time. Like most other classical economists, Malthus emphasized the existence of biophysical limits to growth, remembering the land shortages which a few decades earlier had seemed to constrain England's economic expansion. Ricardo, and later Marx, objected that the development of new technologies, by increasing productivity, would transcend such constraints. Ricardo argued that access to capital and labor could compensate for a shortage of land—this notion that the factors of production are substitutable was crucial to the new approach to economics which he helped to establish, and is still predominant to this day. Marx, too, had confidence in technology and labor, but emphasized that what propelled technological development during the nineteenth century was the desire of the owners of capital to increase their profits, which in his view was done at the expense of the working class. The incentive behind mechanization, in other words, was to lower costs of production, compared to paying wages to a larger and less mechanized work force. By producing and selling a larger volume of products per hour of human labor, capitalists could increase their net income and invest in further technological improvement. Marx also argued that these profits and investments ought to be the collective property of the working people, an argument that profoundly influenced the politics of the twentieth century.

A central question, for our comparative understanding of imperial metabolism, is what Ricardo's and Marx's objections to Malthus really signified, from a global perspective. Ricardo was obviously right in maintaining that the shortage of land would not be an obstacle for England's economic growth, but the technological development which made it possible to transcend the country's biophysical limitations actually implied that England's pressure on the environment was displaced to areas outside its own political boundaries and to future generations yet unborn. In other words, the limits to growth posited by Malthus did not disappear but were shifted beyond view. Even if we disregard the vast quantities of labor time invested in the British colonies to subsidize Britain's economic growth during the nineteenth century, we can join Pomeranz (2000) in calculating the equally vast land areas claimed for the British economy. Already in the mid-eighteenth century, the annual British import of Swedish iron represented around a million hectares of Swedish forest (Warlenius 2011: 68). To substitute for the food energy in sugar consumed in England in 1831, the country would have needed to grow domestic food crops on an additional million hectares of farmland. To replace the cotton fiber imported in 1830 with domestic wool, England would have required an additional 9.3 million hectares of pasture and hay. To replace the annual import of Baltic and American timber in the early nineteenth century would have required almost 0.65 million hectares of British woodland, and to substitute firewood for the annual consumption of coal around 1815, another 6 million hectares of forest. During the course of the nineteenth century (from 1815 to 1900), Pomeranz adds, England's imports of sugar increased elevenfold, its coal output fourteenfold, and its cotton imports twentyfold. In the year 1900, these three commodities alone (sugar, coal, and cotton) thus implied an ecological relief amounting to over 200 million hectares of ecoproductive land. Rolf Peter Sieferle (2001 [1982]: 104) calculates the wood equivalent of British coal extraction in the year 1900 as in itself over 225 million hectares (2,252,000 km^2) of woodland. If we include, in addition to sugar and cotton, other land-intensive imports such as grains, beef, timber, and a variety of colonial crops such as coffee, tea, and tobacco, it becomes apparent that this ecological relief surpassed the *total* land mass of Great Britain (less than 24 million hectares) by at least an entire order of magnitude.

The extraction and transport of these and other imports to England was to a large extent financed with revenue from textile exports. Ultimately, in other words, the point with all the investments in intensified mass production was that it granted England access to increasing volumes of resources beyond its own land surface. This reinterpretation of the Industrial Revolution in terms of global transfers of resources has not taken into account the immense amounts of labor invested in colonial plantations, mines, and forests, or the vast land areas which provided all these laborers with food. In acknowledging the requisite appropriation of labor and land in the periphery, this perspective on industrialization finally leads us to recognize that technology may not primarily be a matter of *saving* time and space, but of *redistributing* it in global society (Hornborg 2006). Fundamental to such asymmetric transfers of embodied time and embodied space in the world-system, of course, are global discrepancies in the *price* of labor and land. Technological rationality, in other words, is a subset of mercantile rationality (what the economists refer to as arbitrage), and technological progress is contingent on global market conjunctures.

Even these cursory calculations can contribute to a reassessment of the essence of technological development in a global perspective. Malthus and the other classical economists were right in concluding that there are limits to the amount of land area that is available to a nation's economy, but Ricardo was also right in observing that England could transcend such limits by substituting capital and labor for land—although, as we have seen, this largely meant shifting its land requirements to other nations. To the extent that recurrent concerns with the "limits to growth" can be justified, whether raised by the threat of environmental degradation, energy scarcity, resource depletion, food shortages, climate change, global inequalities, or financial collapse, it is essential to understand the emergence and expansion of industrial technology as a total *social*—and world-historical—fact.

Continuities and Discontinuities in Imperial Strategies

The empires briefly reviewed above vary enormously in geographical extent, population size, and institutional framework, from the few million inhabitants of the Aztec realm in 1519 to the 650 million citizens of the British Empire four centuries later, but in order to understand the world-historical transformations leading from tributary agrarian states to mercantile and industrial expansionism, we need to address the continuities as well as the discontinuities. To begin with, we can observe that all the imperial projects mentioned have relied, to use the nutshell definition of Jane Burbank and Frederick Cooper (2010: 10), on "the political logic of enrichment through expansion." All empires aim to control the land and labor of a vast territory in order to accumulate wealth or capital in the hands of a powerful elite affiliated with its core. In all the cases we have reviewed, imperial projects can be viewed as attempts to control politically (and coercively) preexisting systems of agricultural surplus production and long-distance exchange, that is, to turn trade into tribute.

As argued above, most of the productive potential of land and labor in an agrarian society ultimately represents solar energy captured through photosynthesis. The mainly agrarian, tributary empires in our sample (Han China, Rome, Inca, and Aztec) relied on a combination of religious devotion and military coercion to channel such energy to make it accessible for elite control. In the final two cases (the Spanish and British empires), we can more clearly see how they were at least initially geared to global market conditions, relying on advantageous exchange rates to complement religion and coercion in guaranteeing the accumulation of wealth. Market institutions were also variously employed in imperial China, Rome, and Mexico (less so among the Inca), but the divinity and military power of the emperor remained the dominant prerequisites to accumulation. The largely mercantile origins of Spanish and British imperialism in no way means that religion and military coercion were dispensable to them, but that global structures of market demand had become indispensable (Wallerstein 1974–1989; Braudel 1992 [1979]). Without the global patterns of demand for silver, sugar, and cotton textiles, such mercantile empires would not have appeared and continued to thrive. For Han China, Rome, Inca Peru, and Aztec Mexico, the preexistent networks of long-distance trade were certainly a condition for their appearance, but once imperial power was established there was a tendency for commerce to become suppressed and secondary to administered exchange.

If the acceptance and use of market institutions prompts us to draw a distinction between those imperial expansions that are more or less dependent on such institutions, and those that are not, we also need to consider the distinction between primarily mercantile empires (such as the Spanish in the sixteenth century) and industrial empires (such as the British became in the nineteenth century). As suggested above, the industrial strategy of accumulation can be visualized as a subset of the mercantile strategy, in that both rely on global discrepancies in the market prices of labor and land-based resources. It is trivial to reiterate the economic rationality of, for instance, the spice trade, in which the opportunity for accumulation hinged on the huge differences in the market price of spices in Indonesia and Europe. It is considerably less trivial, however, to observe that for eighteenth-century Europeans to find it rational to replace local labor with mechanical devices requiring the input of labor and natural resources from some remote periphery (such as forests and iron mines in Sweden, or coal mines in the northwest of Britain), the relative price of local versus peripheral labor should be a crucial consideration. By way of concrete illustration, we might imagine how different the conditions for industrialization in nineteenth-century Britain would have been, if the labor that harvested the cotton fiber in the colonies had been paid standard British wages, and the land rent for colonial plantations had been equivalent to that of prime British farmland. From this perspective, the mechanization of production is inextricably intertwined with market conditions, as part and parcel of a strategy of mercantile conversion. The very existence of modern technology, in other words, relies on specific rates of exchange (of embodied labor time and natural space) between different sectors of world society.

Having reconceptualized industrialization as an imperial strategy comparable to accumulation through tribute or mercantile capitalism, we next ought to reconsider the concept of capital itself, as well as the ideologies that tend to accompany and legitimize these various modes of capital accumulation. In a comparative, world-historical perspective, capital accumulation can be understood as a recursive (self-reinforcing) relation between some kind of material infrastructure and a symbolic or coercive capacity to make claims on other people's labor or land-based resources. This very general definition of capital would thus include, for instance, not only the specifically "capitalist" relation between nineteenth-century British textile factories and the institutions of wage labor and market exchange, but also the similarly recursive relation between sixteenth-century agricultural terraces in Peru and the Inca institution through which maize beer was ceremonially redistributed in exchange for labor. With this definition, both the textile factories and the terraces qualify as capital in the sense of a material infrastructure that is accumulated through a specific cultural strategy for appropriating a net social transfer of labor and resources, and the expansion of which enables an expanded appropriation of such transfers in the future. In the sample of empires reviewed above, we have recognized as examples of such accumulated capital a range of investments including agricultural infrastructures, roads, canals, ships, armies, ceremonial architecture, factories, and railroads.

Moreover, accumulation presupposes objectively quantifiable rates of unequal exchange that guarantee net transfers of labor time or natural resources from one segment of society to another. Following Godelier (1986), however, we expect such asymmetries in exchange to be systematically mystified and instead presented as reciprocal or fair. Using our comparison above, we can thus juxtapose Inca concepts relating to the redistribution of maize beer (for instance, the *minka* labor parties) with modern concepts of "wages" and "market prices" as comparable instances of mystification. Other examples of cultural ideologies that have represented appropriation as reciprocity would include beliefs concerning the various divine services of emperors in Han China, ancient Rome, and the Habsburg version of the Holy Roman Empire.

Ecological Imperialism Redefined: Imperial Strategies of Ecologically Unequal Exchange

Although more or less intuitively based notions of unequal exchange or exploitation are common in the social-science literature, particularly within the paradigm of world-systems analysis, attempts to analytically define such concepts tend to be flawed and confusing (Hornborg 1998, 2013). However, if empires through world history have indeed been oriented toward "enrichment through expansion," there is every reason to carefully consider if and how they have been engaged in unequal exchange with their conquered territories.

Predictably, the concept of unequal exchange makes little sense to mainstream economists, whose focus on monetary exchange values (prices) generally implies inattention to other metrics such as energy, materials, embodied labor time, or embodied land, by means of which an inequality of exchange might be assessed. Similarly, because economists tend to understand capital as abstract monetary wealth rather than material infrastructure, it is difficult for them to see the relevance of positing an unequal exchange of material resources (including labor, which is actually a form of energy) that contributes to the accumulation and maintenance of capital. Discussions of unequal exchange have thus generally been confined to schools of thought concerned with finding nonmonetary perspectives on resource flows, primarily Marxist and ecological economics (Emmanuel 1972; Bunker 1985; Lonergan 1988;

Odum 1996). In these discussions, unequal exchange has generally been conceptualized as the deviation of prices from "values" in international trade, generating an asymmetric (net) transfer of value between different segments of society (generally nations). The underpaid value has been defined in terms of either embodied labor or embodied energy. Such materialist definitions of value, however, are difficult to reconcile with what social and cultural theory has to say about the semiotics of consumption. Nor are they conducive to constructive discussions with mainstream economists, who also tend to be concerned with what consumers actually consider valuable. For these and other reasons, it is analytically more valid to decisively distinguish definitions of unequal exchange from considerations of value (see chapter 5). The asymmetric transfer of various kinds of resources—unequal exchange of energy, materials, embodied labor time, or embodied land—is indeed crucial for the accumulation of capital as defined above, but the assessment of such asymmetries should not be geared to notions of underpaid value. Keeping physical resources and economic value apart seems to be the only analytically tenable way to proceed in reconciling the interaction between material and semiotic aspects of economic processes.

These considerations clearly have relevance for the study of largely mercantile empires such as the Portuguese, Dutch, Spanish, and British, where particular constellations of consumer preferences were intertwined with material processes of capital accumulation. In other words, the transoceanic trade in goods such as spices, sugar, tea, and beaver pelts yielded monetary profits that could be invested in more ships, armies, and, ultimately, factories. The accumulation of such infrastructures through market transactions clearly entailed asymmetric flows of timber, foodstuffs, ores, and other resources that could be approximated in terms of unequal exchange, but to approach such flows in terms of underpaid values, dissociated from the actual preferences of market actors, would lead nowhere. To demonstrate that a given pattern of market transactions entails a systematically unequal exchange of embodied land or labor time can help to account for the accumulation of capital, but rather than proposing that land or labor is underpaid, it will suffice to show that market prices function as an ideology of reciprocity that mystifies such asymmetries.

In turning from mercantile to ancient tributary empires with little or no market institutions, we need to ask to what extent the accumulation of capital can be viewed as the product of unequal exchange: can labor and resources appropriated through tribute at all be said to be *exchanged*? Although concepts of unequal exchange have been developed to expose asymmetries in trade masked as market reciprocity, the underlying and wider notion of exploitative, net transfers of resources must obviously include tribute as well. In several cases (for instance, the Aztecs), the distinction between trade and tribute is in fact difficult to draw. Most forms of tribute, even in as nonmercantile an empire as that of the Inca, are *conceived* by the tribute payers as a kind of exchange for services provided by the emperor. Such a perceived exchange of services can generally be assessed in terms of objectively quantifiable flows of embodied labor and land. Thus, for instance, it would be fairly simple to demonstrate that the maize beer which the Inca emperor served his subjects only represented a fraction of the harvest that he gained from their labor. Any kind of societal flows of goods or services, including tribute, should be considered part of an *exchange*, and market exchange is simply one among several institutions—and market vocabulary one among several ideologies—for organizing unequal exchange so as to render it invisible.

The Roles of Money and Technology in the Ecology of Empire

If much economic and technological expansion can indeed be visualized as the product of ecological imperialism, redefined as ecologically unequal exchange, we must conclude that mainstream beliefs about the societal roles of money and technology are incomplete. The gradual world-historical transition from tributary to increasingly mercantile and industrial imperialism is generally viewed as progress, and in some respects this is certainly a valid perspective. At the same time, we must not ignore the global inequalities and unevenly distributed ecological consequences that have accompanied this development, which might justify the question, "Progress for *whom?*" Whatever transformations the next few centuries have in store for world society, whether managed or unintentional, it must be important to develop as full as possible an understanding of global socioecological processes. This should include recognizing that market exchange can mask asymmetries in global resource transfers, and that what we think of as technological development can in fact rely on such asymmetric transfers. Even globalized environmental problems such as climate change have highly unequal repercussions for different parts of global society (Roberts and Parks 2007). Moreover, we need to acknowledge the temporal dimension by considering how the combustion of fossil fuels and depletion of finite resources also imply a colonization of future generations everywhere.

There is no doubt that general-purpose money and global market exchange provided the means for new forms of ecological imperialism, particularly after the establishment of intensive transoceanic trade in the sixteenth century. As we saw in chapter 3, money is a curious cultural institution that continues to mystify socioecological processes such as the accumulation of new technologies in particular sectors of the world-system. Toward the end of the twentieth century, the idea and institution of money became the foundation of yet another and even more opaque strategy of accumulation that we might call *financial* imperialism (cf. Graeber 2011a). Much as in earlier imperial strategies of enrichment, our difficulties in grasping its exploitative nature is an essential condition for its existence. Power is founded on obscuring the relation between the material and the symbolic. Ubiquitously, it entails both unequal access to material resources such as energy or land *and* cultural mystification of such inequalities (for instance, through concepts such as "market prices" or "debt"). Ubiquitously also—not only in Marx's nineteenth-century capitalism—political relations with other people masquerade as relations to things. These things—artifacts in the sense of human agency interlaced with matter—include landscapes, commodities, money, and technology.

Environmental historians have assembled the details of landscape transformations in the wake of imperial expansions. In the ancient agrarian civilizations, the most conspicuous cultural transformations of landscape were deforestation and the establishment of various kinds of landesque capital (Williams 2003, 2007; Widgren 2007; Håkansson and Widgren 2014). The traditional Chinese landscape of rice paddies, mulberry trees, and fish ponds was largely a product of two millennia of Chinese Empire, just as the Mediterranean landscape of grape vines and olive trees was a product of ancient Greece and Rome. The terraced Andean slopes and irrigated coastal valleys of Peru likewise reflect millennia of state expansions, ending with the Inca, as do the extensive investments in raised fields in waterlogged areas such as the basins of Titicaca and Mexico. The landscape legacies of the Spanish and British empires include abandonment of indigenous landesque capital and an initial reforestation in the wake of indigenous depopulation, followed by renewed and unprecedented deforestation and cultivation in connection with colonization and the establishment of plantations, frequently leading to serious soil erosion and degradation. European expansion in the New World also brought devastated mining landscapes, collapsed fisheries, and a severe decimation of numerous wild species ranging from North American beavers (and the concomitant hydrological and vegetation changes following the abandonment of beaver dams) to Caribbean sea turtles. To this we should add, of course, the introduction and often explosive expansion of Old World species of animals and plants, both wild and domesticated. As Crosby and others have shown, the expansion of European empires since the sixteenth century is largely responsible for the global distribution not only of cattle, sheep, pigs, and wheat, but also of rats and dandelions. These early modern empires may have transformed their natural environments more dramatically than ancient empires such as Rome (cf. Horden and Purcell 2000; Woolf 2012: 56–61), but it is no exaggeration to observe that significant proportions of landscapes on all continents have been molded by the world-systemic conjunctures of imperial command and market demand over the past two millennia. These landscapes can thus be viewed as the historical imprints of social and political relations between humans inscribed in the ecosystems of which they are a part.

The goods that constituted the essential metabolic flows in the empires mentioned above, and the production of which in large part shaped their landscapes, are in themselves perhaps the clearest illustration of how relations between people masquerade as relations between things. Commodities are ultimately embodiments of human labor and natural resources, but present themselves to our senses as decontextualized items of exchange. Their exchange on the market, including the rates at which they are exchanged, has no systematic relation to the quantities of labor time or natural resources that they embody. The extent to which their exchange entails asymmetric transfers of embodied labor time or natural space must thus be investigated separately, without guidance from the vocabulary of economics, although the exchange rates help us to empirically assess the specific quantities of labor time or natural space that are transferred in a particular context.

Like commodities, money was recognized by Marx to have a fetish-like aspect, in that it focuses our attention on a concrete reification of a wider social system of material exchanges. Money fetishism prompts us to attribute to nonliving objects like bills and coins certain properties of living things, such as a capacity for autonomous growth (i.e., interest), rather than acknowledging that the appearance of having such properties is a result of social relations of exchange. As argued above, this capacity of money to mystify the material substance of unequal exchange has for a very long time been an important ingredient in imperial expansion. The literature attempting to grasp the emergence of this elusive cultural phenomenon is vast and impossible to summarize, but a very useful attempt is the comparative economic anthropology of David Graeber (2011a). Although his analysis is not concerned with material asymmetries in resource flows, or the material dimensions of capital accumulation, he persuasively relates the world-historical emergence of money and markets to cultural conceptions about predatory lending, debt slavery, state institutions, mercenary soldiers, and violence. Money, we suggested above, is basically a symbolic capacity to make claims on other people's labor and landscapes. It is an idea, embodied in artifacts, which developed as a means of extending control over people beyond relations of kinship, proximity, and trust, in order to get them to behave in ways they might not otherwise have done. Like other forms of artifact-mediated sociality, the facelessness of most commercial transactions permits market actors to develop a kind of psychological dissociation vis-à-vis the social and ecological consequences of their actions.

Finally, as argued above, technology is also a category of artifacts which can mystify unequal relations of exchange. Viewed from the perspective of time–space appropriation, technological objects reflect their owners' harnessing of the deflected agency of other people. Modern technology can be reconceptualized as a strategy to locally save (human) time and (natural) space, at the expense of time and space lost elsewhere in the world-system. Although it has been suggested that the earliest proto-machines (water mills) were built in the late Roman Empire in part to replace increasingly expensive slaves—as a continued preoccupation with the *delegation* of work to other beings/things— we return to the fundamental question if technology has not so much *re*placed as *dis*placed slavery. We need only to think of the role of Caribbean plantation workers and British coal miners during the Industrial Revolution, or the working conditions of Peruvian copper miners and Brazilian sugarcane harvesters today. Like commodities, technological objects are fetishized, obscuring asymmetric exchange relations and distant investments of labor time, embodied land, or energy. The Western fetishization and glorification of technology has been abundantly documented, particularly by historians of British and American imperialism (Headrick 1981, 1988, 2010; Adas 1989, 2006; Marsden and Smith 2005). It is a pervasive component in the mainstream narrative of development, which legitimizes the hegemony of wealthier nations by representing poverty as failure and global inequalities as a temporary state of affairs.

Ecological Imperialism Today and Tomorrow

In considering the prospects for a comparative study of imperial metabolism, we have suggested that the appropriation of human time and natural space, in the form of embodied labor and embodied land, might serve as quantifiable parameters that transcend specific production systems and energy regimes. Labor and land, always intertwined, are indeed universal sources of energy for capital accumulation. To understand current trends in ecological imperialism, we need to fundamentally rethink the rationale of the Industrial Revolution.

Prior to the Industrial Revolution, requirements for energy and land converged in the production of food for human labor and fodder for draft animals, but for the past two centuries, fossil fuels have made it possible for some sectors of the world-system to separate their energy and land requirements. This material condition was the foundation of the industrial worldview that emerged in early nineteenth-century Britain, the "image of unlimited good" (Hornborg 1992; Trawick and Hornborg 2015) that pervades mainstream economic thought to this day. Faced with twenty-first-century prospects of peak oil and climate change, this worldview is now being more seriously challenged than ever. We need to ask if some of the fundamental tenets of modern economics are inextricably connected to the use of fossil-fuel energy. A return to biofuels would reintroduce some of the constraints and rationalities of preindustrial imperialism, including the ancient competition over land for energy versus food production, and the logic of calculating transport costs in terms of the requisite ecological space.

Although sharing much continuity, nineteenth-century British and twenty-first-century American imperialism are in some respects diametrical opposites. Britain's imperial strategy was largely propelled by its great need of additional land, while it was more than self-sufficient in fossil fuels. The contemporary United States is apparently prepared to go to war to secure its imports of fossil fuels, but is more than self-sufficient in agricultural land. If, prompted by peak oil or global warming, net oil imports to the United States would have to be replaced with best-practice Brazilian ethanol, it would require 187 million hectares of sugarcane plantations, which is more than seven times the area within the United States currently devoted to agricultural exports and almost 47 times the area in Brazil currently devoted to sugarcane ethanol. Without a doubt, the future of ecological imperialism and our understandings of ecologically unequal exchange will, to a large extent, hinge on the geopolitics of energy.

Contrary to many theorists contemplating the expansion of the West, we should not let this acknowledgment of the significance of energy in the world history of empire lead us to adopt a simplistic, Darwinian perspective representing the hegemony of high-energy and high-tech imperialism as natural and inevitable (cf. Crosby 2006). Such evolutionary determinism is pervasive in the work of cultural ecologists like Leslie White (1959) and recurs, for example, in the influential narrative of the biogeographer Jared Diamond (1997). References to the relative amount of energy harnessed or controlled per capita, however, do not suffice to explain the world history of imperialism. Access to powerful technologies may certainly be useful to empires, best illustrated by British and American imperialism in recent centuries, but the complex vicissitudes of imperial fortunes are equally dependent, for instance, on the cultural vagaries of global markets, political intrigues, epidemiology, harvests, and even weather (recall the fate of the Spanish Armada). Empires rise and fall, and in many historical instances low-energy and low-tech polities have proven more resilient than their more powerful neighbors or predecessors.

CHAPTER 5

Money as Fictive Energy: Unraveling the Relation between Economics and Physics

On April 8, 1880, the Ukrainian physician Sergei Podolinsky (1850–1891) wrote a letter to Marx, asking for his opinion on a text he had sent him a week earlier, in which he attempted to integrate Marx's theory of surplus labor value and thermodynamics. Podolinsky had shown that humans are "perfect machines" in the sense that their labor, particularly in agriculture, can enhance the accumulation of available energy on Earth. He was not the first to argue that humans and draft animals are more efficient energy converters than steam engines, but he was the first to try to connect a labor theory of value with an energy theory of value (Martinez-Alier 1987: 51). We do not know what Marx himself may have thought about the idea, as there are no records of his response, but Marx's close friend and colleague Friedrich Engels dismissed Podolinsky's reasoning. In a letter to Marx in 1882, Engels wrote that "what Podolinsky has completely forgotten is that the working man is not only a fixer of present solar energy, but more than that, a squanderer of past solar heat."

The failure of this potential dialogue between Marxian and ecological economics is emblematic of the pervasive historical disjunction between concerns pertaining to the management of money and energy, respectively. It is particularly frustrating and revealing because Marxian economics presents itself as based on historical materialism, much of the Marxian framework suggests an intuitive concern with thermodynamics, and both Marxian and ecological economics pose influential, "heterodox" challenges to mainstream economics. Podolinsky had certainly not "forgotten" that industrial production processes were based on squandering finite stores of ancient solar energy, but the most notable aspect of Engels's response is the extent to which his own thinking about economic processes was entrenched in the industrial worldview based on the use of fossil energy. Throughout the past two centuries, the widespread intuition that the industrial world order is an unsustainable distortion of human–environmental relations has been based on an understanding of the role of humans in the biosphere *as if they did not have access to fossil fuels*, while the various arguments in defense of industrialism— even from some of its most heterodox critics—have been based on the contrary logic imposed by fossil fuels.

Money as Fictive Energy

Although the concept of energy did not appear until the nineteenth century—and the laws of thermodynamics were formulated to theoretically grasp the problem of energy efficiency in steam engines—humans appear to have had various corresponding notions for millennia. As noted in chapter 2, preindustrial societies, whether based primarily on hunting and gathering or agriculture, were no doubt aware of the fact that solar energy is the vital flow that ultimately animates all life on Earth, including wild plants, game animals, crops, draft animals, and human beings. With expanding commerce and money use, however, some social groups, particularly those specialized in trade, were able to view flows of money as more fundamental to their subsistence than flows of solar energy. The expanding commerce was thus accompanied by various cultural representations of money, of which the worldview of mainstream economics is an example (cf. Gudeman 1986). The conceptual framework of neoclassical economics is in part inspired by the nineteenth-century European preoccupation with developing the concept of energy, and vice versa (Mirowski 1989). Significantly, however, the understanding of energy that may have inspired nineteenth-century economists was based on the First Law of Thermodynamics, but not the Second. To the extent that economics was inspired by physics, in other words, it failed to be concerned with the irreversible degradation of energy. This fundamental difference between physics and economics has crucial implications (Glucina and Mayumi 2010: 22).

The history of the relation between notions of energy and notions of monetary value is deeply paradoxical. Energy and economic value are both decontextualized, quantifiable abstractions developed to understand and manage the operation of the natural and social systems in which humans participate. Even if the sciences of physics and economics have common roots, a recurrent criticism of mainstream economics throughout the nineteenth and twentieth centuries has been that it ignores physical flows such as energy (Martinez-Alier 1987). There have been many attempts, frequently by people with a background in natural science, to rewrite the science of economics from the perspective of an energy theory of value (Mirowski 1988). The response of neoclassical economics has generally been a conspiracy of silence (ibid.: 818), but rather than dismiss the relevance of energy for economic theory, it is reasonable to expect the discipline and profession of economics to seriously consider, as did Georgescu-Roegen (1971), the implications for economics of the laws of thermodynamics. Although completely alien to the current preoccupations of mainstream economics, it would reintroduce a concern with material aspects that, as we saw in chapter 3, has been expelled from economics since the deliberations of classical political economy.

Such a concern would undoubtedly require a fundamental reconceptualization, within neoclassical economics, of the prospects of economic growth (Daly 1996; Hamilton 2003; Victor 2008; Jackson 2009; Glucina and Mayumi 2010). It would seem particularly relevant for economics to investigate the possible connections between financial crises and the declining net energy or EROI (Energy Return On energy Investment) in modern production processes (Hall and Klitgaard 2011; Lipson 2011), particularly as the phenomenon of diminishing returns has been posited as a recurrent cause of large-scale societal collapse through history (cf. Tainter 1988).[1] The strong historical correlation between economic growth and energy consumption suggests a causal link, and despite widespread claims that it is feasible to "delink" or "decouple" them, it has not been demonstrated (Jackson 2009: 67–86; Glucina and Mayumi 2010: 17–19).

As indicated in chapter 3, a renewed concern with the material aspects of economic systems might also be conducive to renewing a concern, within economics, with morality. The foundation of both Marxian and ecological theories of unequal exchange is the recognition that capital accumulation is based on asymmetric net transfers of biophysical resources such as embodied labor, land, energy, or materials (Emmanuel 1972; Bunker 1985; Lonergan 1988; Odum 1996; Hornborg 1998). Although we shall show that it is misleading to equate such embodied resources with economic value—as in labor or energy theories of value, which reproduce the old confusion between energy and money—the various deliberations on the discrepancies between them could add up to a theory of unequal or asymmetric exchange which, though in itself objective and nonnormative, raises moral questions. Common to all these deliberations is the understanding that market prices, to which the interests of economists tend to be exclusively confined, project an illusion of reciprocal exchange that conceals the asymmetric material transfers that are prerequisite to accumulation but are beyond the horizons of mainstream economic thought. The market mechanisms and ideological blinders that allow such asymmetric transfers to continue can be identified as readily as the actual transfers themselves (for the latter, see Erb et al. 2009; Dittrich and Bringezu 2010; Duchin and Levine 2012; Lenzen et al. 2012, 2013; Yu et al. 2013; Alsamawi et al. 2014; Dorninger and Hornborg 2015; Simas et al. 2015), but the role of these material transfers in reproducing uneven global patterns of accumulation and development cannot fail to raise moral issues that remain invisible for economists preoccupied with the intricacies of market equilibrium.

The abandonment, within mainstream economics, of concerns with the material aspects of economies has prompted voluminous protests from the start. The most elaborate and conspicuous of these dissident approaches are those of Marxian and ecological economics. Although there have been attempts by so-called eco-Marxists to reconcile these two approaches, significantly by conceding that the Marxian concept of labor-power is cognate to, or even a type of, energy (Burkett 2005b; Foster and Holleman 2014; Moore 2015),[2] there are differences that are difficult to straddle.[3] While the two traditions share the conviction that market prices, or exchange values, do not do justice to "use values," and that the underpayment and unequal exchange of such use values are fundamentally problematic, they differ in terms of which use values are central to their analyses and in terms of what the central problems are. The Marxian framework traditionally focuses on the industrialists' appropriation of the use value of labor, and on the resultant inequalities and polarization of social classes. The outlook of ecological economics, on the other hand, has focused on the appropriation of "natural" use values such as energy, embodied land, and ecosystem services, and on the resultant degradation of the natural environment.[4] For a synthesis of these two critiques of industrial capitalism to progress, the strong Marxian emphasis on its labor theory of value would need to be reconsidered. Even from a Marxian perspective, the unique significance attributed to labor-power, in relation to all other inputs in production, is an analytically flawed position (Keen 1993; Brennan 2000).[5] Furthermore, the conceptualization, by both schools, of biophysical resources—whether human labor, energy, or land—as underpaid "use values" is a misleading way of addressing the material aspects of the economy. Paradoxically, this understanding of unequal exchange in terms of underpayment is fundamentally similar to the conviction of neoclassical economics that market prices may not do justice to externalities, and that the challenge is to internalize insufficiently compensated costs such as damages to the environment or to human health.[6] To criticize capitalism by attempting to redefine economic value—a concept invented by merchants—is not an effective approach.[7]

The notion of underpayment is misleading in several ways: (1) If the appropriated resources were to be fully compensated for their contribution to the value of the finished product, the incentive to conduct industrial production—that is, the possibility of profit—would diminish, as the rationale of industrial capitalism is precisely *not* to compensate labor and land for the exploitation of their productive capacities. (2) Moreover, if "use values" are biophysical resources such as labor, energy, and land, they cannot be quantified in monetary measures, which means that it would be difficult to argue that they are underpaid, unless laborers are malnourished or ecosystems collapse, that is, when they are not provided with the monetary means of sustaining their basic metabolism. (3) Finally, ever since Aristotle, the concept of use value has referred to the use to which a product can be put in satisfying a person's needs, but the use value of most modern commodities is largely or even entirely determined by symbolic factors, rather than by the volume of biophysical resources that they represent.

To resolve these contradictions, materialist critiques of industrial capitalism would need to abandon notions of underpaid use values in favor of the incontrovertible observation that capital accumulation requires asymmetric transfers and the irreversible degradation of biophysical resources. The asymmetric transfer, or appropriation, of biophysical resources such as embodied labor, energy, land, and materials is certainly orchestrated by market prices, or exchange values, but it only confuses matters to propose that those biophysical resources are in reality more authentic measures of value than the exchange values experienced by market actors. Moral indignation over the tendency of economic logic to generate social polarization and environmental degradation is entirely appropriate, but to express such indignation in terms of underpayment is paradoxically to subscribe to the faith in a common metric of value which underlies the ideology of the market. It is not useful to claim that economic value is something else than what is established through market exchange. The insistence, by many ecological and most Marxian economists, that this is indeed the case, presents one of the greatest obstacles to dialogue between heterodox and mainstream economics.

Although the contradictions between heterodox and mainstream economics may seem fundamental and insurmountable, all these various schools are united by a common trust in the phenomenon of general-purpose money. Even if their political recommendations appear to be very far apart, they share the assumption that problems can be alleviated without questioning money itself. However, if the analyses and conclusions of Nicholas Georgescu-Roegen (1971) are taken seriously, neither proposals for economic redistribution— whether through price changes, taxation, subsidies, or a more radical reform of ownership—nor technological progress will solve the fundamental problems of sustainability confronting market-based economic processes orchestrated by general-purpose money in a universe obeying the Second Law of

Thermodynamics. In pricing commodities representing dissipated resources higher than those resources contemporary forms of money and market exchange will inexorably reward an accelerating dissipation of resources. The problematic relationship between general-purpose money and thermodynamics inevitably also generates unequal exchange in the sense of objectively asymmetric transfers of biophysical resources from extractive sectors to core regions of the world-system.

The societal implications of these material polarizations, asymmetries, and inequalities have been conceptualized in transdisciplinary theory unraveling the connections between thermodynamics and political economy (Bunker 1985; Hornborg 1992, 2001a; Biel 2006, 2012). The uneven accumulation of technological infrastructure in core sectors of the world-system has been made possible by the interaction of money, prices, and the exchange of biophysical resources over the past three centuries. From the days of the steam engine, the conditions for technological progress were no longer restricted to human ingenuity, but *required* substantial differences in the prices of labor and resources between different parts of the world-system. This intertwining of global societal exchange rates and local material productivity has remained beyond the horizons of economic thought. The inability of mainstream economics to perceive the material determinants of uneven development and environmental degradation continues to preclude effective solutions to these challenges of sustainability. Our focus in this chapter is how, since the adoption of steam power, even heterodox approaches to economics have been constrained by the worldview promoted by the use of technologies propelled by fossil energy.

"Use Values," Underpayment, and Unequal Exchange

In two articles expressing strong faith in technological progress, the Marxist biologist David Schwartzman (1996, 2008) criticizes Georgescu-Roegen's (1971) analysis of the relation between thermodynamics and economics and his pessimism regarding the potential of solar energy. Schwartzman (2008) also criticizes Paul Burkett's (2005a) efforts to reconcile Marxist theory and ecological economics. Burkett basically agrees with Georgescu-Roegen that the dissipation of matter-energy and the practical impossibility of complete recycling pose significant constraints on human economies. Although Burkett's ambition to incorporate ecological concerns into Marxist theory is laudable, Schwartzman's objections are consistent with the outlook of classical Marxism. The Promethean strain in Marxist thought (Benton 1989) cannot be denied. Marx's concern about the depletion of rural soils (Foster 2000) did not make him any less optimistic about the prospects of fossil-fuel technologies. It is true that Odum's (1996) concept of emergy is analytically identical to Marx's concept of labor value, but this does not make Marx an ecologist.

The divergent approaches to the prospects of technological progress evident in Marx and Georgescu-Roegen, respectively, highlight a division between two divergent strands of ecological economics: one which duplicates the Marxian faith in the possibility of reconciling industrialism and sustainability,[8] and another (following Georgescu-Roegen) which offers a more profound and radical rethinking of economics, the implications of which have not yet been fully fathomed. The former approach views technological capital as an accumulation of embodied labor and other energy which *in itself* is exempt from political criticism and can be put to morally defensible use, whereas the second approach ultimately implies that uneven technological growth must be viewed as an index of unsustainable resource dissipation and unequal exchange.

Burkett's efforts are significant and instructive, for they reveal divergent assumptions about economic processes that ultimately cannot be reconciled. In an earlier article, Burkett (2003: 138–141) compares the perspectives of ecological economics, Physiocracy, and Marxism regarding the sense in which nature can be considered a source of economic value. While many ecological economists treat "nature as a direct source and substance of value" (for instance, Costanza 1980; Odum 1988), others (Georgescu-Roegen 1971; Daly 1996) are content with observing that natural resources—sources of "low-entropy matter-energy"—are consumed and dissipated in the production of valuable goods and services, while the definition of the latter as valuable is based on immaterial gratification contributing to what Georgescu-Roegen referred to as "psychic income" and the "enjoyment of life."

The difference between these two schools of thought is important. The former approach offers a physical theory of value, in effect equating economic value with quantifiable, past investments of some material resource. Because of the continuities clearly linking it to eighteenth-century Physiocracy, I refer to it as "neo-Physiocrat" ecological economics. The latter approach actually accepts the mainstream perception among neoclassical economists of consumer utility as equivalent to economic value, reflected in market price, while adding the crucial observation that the production of economic value simultaneously increases entropy and environmental degradation. This analytical distinction between the largely cultural dimension of consumer or market value (Burkett's "exchange value") and the material dimension of physical resource theory (misleadingly referred to as "use value") is essential to any attempt to reconcile the interaction between semiotic and material aspects of economic processes. Rather than reducing economic value to embodied quantities of a physical force or flow, it makes it possible to show how these two phenomena are related to each other.

Although indicating the same distinction, the concepts "cultural" versus "material" are more to the point than Burkett's use of "exchange value" versus "use value." Burkett (2003) suggests that the Physiocrats' focus on material aspects was actually a focus on use value, neglecting exchange value, but does not acknowledge that even use value is defined by the cultural preferences of the consumer. The distinction between exchange value and use value goes back to Aristotle and the monetarized economy of ancient Greece. In Marxist thought, use values are equated with "real wealth" and denote material quantities of resources such as embodied labor, energy, land, and water, which provide for human needs. As Baudrillard (1981 [1972]) and Sahlins (1976) have shown, however, human needs—beyond the bare metabolic requisites of keeping an isolated human organism alive—are impossible to extricate from their cultural context.[9] To illustrate this point, we might ask, how much "use value" pork has for a Muslim? If use values are culturally determined, it is difficult to see how they could be objectively identified with material quantities.

The comparison between Marxism and Physiocracy—including its revival in some strands of ecological economics—is illuminating. Whereas the Physiocrats had perceived land as the ultimate generator of economic value and growth, Smith, Ricardo, and Marx emphasized labor, but the structure of the argument is very similar. All sought to identify a factor of production with the special quality of being able to yield more value than is required for its maintenance. Malthus, too, referred to that special "quality of the earth by which it can be made to yield a greater portion of the necessaries of life than is required for the maintenance of the persons employed on the land" (quoted by Benton 1989: 61, reference to Ricardo).

The structural similarities between Marxism and Physiocracy—including "neo-Physiocrat" ecological economics—are also revealed in Lonergan's (1988) analysis of theories of unequal exchange. Both Marxists and ecological economists tend to understand unequal exchange as the deviation of market prices from "real" values. Whether these values are defined in terms of labor or energy theories of value, Lonergan concludes, the methods and models used are almost identical. Burkett (2003: 139) observes that the embodied energy theory of value "closely and consciously parallels the Ricardian labor-embodied theory of value, with energy replacing labor as the primary factor of production." Similar congruities have recently been identified between the Marxian concept of embodied labor and H. T. Odum's (1988) notion of embodied energy (Foster and Holleman 2014). As Rabinbach (1990) has reminded us, labor is in fact a form of energy.

Although Lonergan does not criticize these theorists of unequal exchange for confusing monetary and material aspects, it is noteworthy that the net labor "value" that Arghiri Emmanuel (1972) and Samir Amin (1976) identify as having been transferred from less to more developed regions in the 1960s is measured in dollars (cf. Lonergan 1988: 135). If surplus labor value is measurable in dollars, it should suffice to conclude that, for a successful capitalist, the price of labor, or other forms of energy, is cheaper than the price of its products. There is no need, in other words, to ascribe to labor or energy a unique role in the creation of surplus value (cf. Martinez-Alier and Naredo 1982: 219; Burkett 2003: 139). Consequently, it should be possible to acknowledge the exploitation of labor without subscribing to a labor theory of value, as well as to acknowledge the unequal exchange of embodied energy without subscribing to an energy theory of value. Moral and political indignation buttressed by theories of unequal exchange and exploitation, in other words, do not require the word "value."

For those of us who hope to strengthen the radical critique of industrial capitalism—and of economic theories that ignore the biophysical dimensions of the economy—that was so powerfully inaugurated by Marx, it is counterproductive to try to cover up for his analytical shortcomings through selective exegesis. Instead, we should be prepared to jettison those aspects of his analysis that were inconsistent with the thrust of his understanding of the logic of capitalism and that can clearly be attributed to the constraints of the hegemonic discourse of his historical context: first and foremost, his Promethean trust in technological progress and his commitment to a labor theory of value.[10]

As we have seen, the conventional understanding of unequal exchange in both Marxist and ecological economics is in terms of underpayment or undercompensation: flows of money or exchange values are represented as not matching the flows of "real" resources, conceptualized as "use values."[11] The implicit assumption is that use values have a "real" monetary value that can be contrasted to actual market prices. However, it cannot be valid to quantify what Marxists refer to as use values, identified with biophysical resources, in monetary terms, as if they had a correct price. It is thus analytically flawed to posit that they are underpaid or undercompensated. It is noteworthy, given the thermodynamic definition of labor-power emerging from Foster and Burkett's (2008) reading of Marx, that this conclusion should be extended even to labor. The concept of use value is in several respects a misnomer, as it can neither be quantified in other than physical metrics nor—as we have seen—extricated, as corresponding to a pure, metabolic need, from the cultural context. The existence of modern technology—the material form of capital accumulation—is certainly predicated on the discrepancy between flows of money and flows of matter-energy, but it is analytically misleading to phrase this discrepancy in terms of an underpayment of use values.

If unequal exchange is instead conceptualized as an asymmetric net transfer of material inputs in production, rather than in terms of an underpayment of material inputs or an asymmetric transfer of value, it will solve another conundrum that has plagued unequal exchange theory from the start, namely, how some extractive economies are able to thrive, rather than become impoverished. All processes of production and capital accumulation build on net transfers of resources—for instance, from rural to urban areas—but whether the transfers imply impoverishment of a given population depends on circumstances of geography and history. Although it is undeniable that ecologically unequal exchange has implied exploitation of large segments of the world's population for centuries, and continues to do so today, the existence of historically privileged and sparsely populated nations richly endowed with natural resources—for instance, Canada, Australia, Scandinavia, Saudi Arabia—has enabled some extractive zones of the world-system to escape economic impoverishment. This in no way contradicts the definition of unequal exchange offered here.

Materialism, Energy, and Surplus Value

Although ecologically oriented Marxists tend to reject narrowly defined energy theories of value, they hold that there are biophysical values in nature that are exploited in capitalism (Burkett 2003: 140). In contrast to neoclassical economics, both Marxist and ecological economics retain the concern of Physiocracy and classical economics with the physical, material aspects of economic activity. Prior to the Industrial Revolution, this concern focused on the productivity of agricultural land, which the Physiocrats recognized as the source of subsistence for all labor, including nonagricultural labor (ibid.: 143). After the turn to fossil fuels and steam engines—and the articulation of the laws of thermodynamics—what is currently known as ecological economics has increasingly focused on energy as defined by physics. While Marxism has maintained its emphasis on the generative capacity of labor, many of the "neo-Physiocrats" have specified their concern with land as a concern with energy. The latter shift seems a logical consequence of the transition from agrarian to industrial society, that is, from a society deriving its energy resources from horizontal land surfaces to one drawing its energy from vertical shafts through the Earth's crust. Ultimately, the attention of Marxism, Physiocracy, and ecological economics to the physical aspects of economic processes all share a concern with energy, as both labor and land can be expressed as measures of available energy.

Did Marx think of energy as in a specific, quantifiable way implicated in the creation of surplus value? Foster and Burkett (2008: 25) agree with Rabinbach (1990) that "Marx always emphasized the energetic basis of labor power and saw it connected to thermodynamics because labor involved mechanical work." According to these authors, there are phrasings of the Marxian labor theory of value which suggest that it is the excess of "productively expendable energy encapsulated in labor power" over the "caloric quantity of useful work needed to produce the worker's commodified means of subsistence" that "enables the capitalist to extract surplus value from the worker," implying that the worker's sale of his labor time is "an energy subsidy for the capitalist" (Foster and Burkett 2008: 26). The very concept of "labor power," it seems, "arose in part from the new thermodynamics" (ibid.: 29). Burkett and Foster (2006) approvingly quote Marx's statement that "labor-power itself is energy," and refer to Marx's "energy income and expenditure approach to surplus value" (120, 126).

Burkett (2003: 142–150) demonstrates the extent to which Marx sympathized with the concerns of the Physiocrats. He mentions that Turgot in 1770 had referred to the ability of the agricultural laborer to "produce over and above the wages of his labor." Unlike modern economists, including Georgescu-Roegen, neither the Physiocrats nor Marx were content with identifying economic value with the immaterial, psychic "enjoyment of life," but struggled to relate it to "the material basis and substance of human life." Marx thus praised the Physiocrats for conceptualizing value and surplus value in terms not of consumption, but of production, and for analyzing capitalist production in terms of "eternal natural laws of production." However, Marx simultaneously recognized that economic value in industrial capitalism could not simply be reduced to material parameters. In his view, the Physiocrats confused "value with material substance." Instead, Marx famously argued that the labor-power of workers had the ability to produce commodities containing more economic value than their wages. Although surplus production as conceived by the Physiocrats seemed a purely material phenomenon, modeled after the physical processes of agricultural production, it was presented as fundamental to the business of making money—that is, earning rent—from owning land. The attempt to *account for monetary gain in terms of physical processes* recurs in Marx's understanding of surplus production in industrial capitalism. This conflation of the material flows of labor energy and the semiotic flows of exchange values/money pervades the Marxian labor theory of value. It builds on important intuitions about connections between energy flows and economic processes, but ultimately does not clarify the nature of those connections.

Marx acknowledged that surplus production in an agricultural society is easier to conceptualize than in industrial society, primarily because it can be identified without the mediation of monetary measurement, but maintained the ambition to understand industrial capitalist profits using a materialist approach largely inspired by Physiocracy. His struggle to reconcile the material and monetary aspects of the economy resulted in inconsistencies such as the notion that the labor theory of value applies only to capitalist forms of production and not to noncapitalist forms.[12] At times, Marx's understanding of economic value formation strongly echoes that of the Physiocrats, as when he refers to the "naturally originating productivity of labor . . . which of course rests on qualities of its inorganic nature—qualities of the soil, etc." (Marx 1967 [1867], quoted by Burkett 2003: 150). Burkett accounts for "Marx's endorsement of this kernel of truth in Physiocratic doctrine" as based on the point that "without an agricultural surplus, there can be no surplus labor in agriculture and no means of subsistence for nonagricultural workers, hence no surplus value in the economy as a whole."

In order to ascertain the extent to which Marxian economic theory rests on an unclear connection between physics and economics, we can consider the famous formula $M-C-M^1$ and ask ourselves whether it is justified to posit a quantifiable relation between the material production of commodity C and the increase in economic value from M to M^1? For Marx, the surplus value is generated by embodied labor. To H. T. Odum (1988: 1136), recently endorsed by leading eco-Marxist scholar John Bellamy Foster (2013; Foster and Holleman 2014), surplus value is generated by embodied energy. The thrust of Odum's argument is very similar to that of Sergei Podolinsky, whose attempt to persuade Marx and Engels about the energy basis of surplus value has been decisively dismissed in a series of articles by Foster and Burkett.[13] The controversy about the so-called Podolinsky business (Martinez-Alier and Naredo 1982; Martinez-Alier 1987, 2011; Foster and Burkett 2004; Burkett and Foster 2006) has largely concerned the question of whether Marx and Engels were adequately versed in thermodynamics, but the crucial question is why that should have been necessary, as the Marxian labor theory of value calculates in money rather than energy. If Podolinsky was wrong about the derivation of economic value from energy, as Foster and Burkett have argued, it is difficult to see why they should need to attribute a cognate perspective to Marx. To the extent that labor-power is indeed a form of biophysical energy, it is reasonable to argue for an affinity between the ideas of Marx, Podolinsky, and Odum, but Foster's and Burkett's position on Podolinsky is contradictory. They dismiss him as an "energy reductionist" (Burkett and Foster 2006: 116) who "confused the physical with the economic" (ibid.: 137). Against this background, it is difficult to understand why it is so important for them to show that Marx had written that "labor-power itself is energy" (ibid.: 120) and that labor is an "energy subsidy for the capitalist" (Foster and Burkett

2008: 26), explicitly referring to Marx's *"energy income and expenditure approach to surplus value"* (Burkett and Foster 2006: 126; emphasis added). The question ultimately is whether they are arguing that Podolinsky was wrong or that his intervention was superfluous? Were there moments, in fact, when Marx himself "confused the physical with the economic"? A truly materialist account of surplus production cannot avoid implicating physics, as Podolinsky recognized, but precisely in not being able to assimilate this insight, the Marxian theory of surplus value as based on labor revealed itself to be entrenched not only in the *operation* of capitalism, but even in its fundamental analytical categories. What Podolinsky recognized was that Marx's notion that economic value derives from the material agency of labor suggests an intuitive understanding of the role of thermodynamics in economic processes.

Foster and Burkett have carefully sifted through Marx's writings in pursuit of every indication of ecological awareness. The quotations they have retrieved are an invaluable distillation of the extent of Marx's orientation in natural science, and their impressive exegetical efforts do not need to be duplicated. What these quotes and Foster's and Burkett's commentaries inadvertently reveal, however, is a pervasive inconsistency in the Marxian framework. A fundamental flaw in Marxian economics is the notion inherited from classical economics, but abandoned in neoclassical economics, that capitalist profits have a specifiable relation—that is, are proportional to—inputs of one particular production cost, namely, labor. The labor theory of value is a survival, within Marxism, of nineteenth-century economics. It has been refuted by virtually all mainstream economists and even some Marxists (cf. Keen 1993) but continues to be taken for granted by most Marxists, not as Marx's own conviction regarding the generative power of labor in general, but as his understanding of the specific mode of operation of the capitalist economy.

However, the notion that labor is underpaid in relation to its contribution to the market price of commodities is an analytically flawed argument. Firstly, the quantity of embodied labor is not measurable in money, only in time or energy, and there is thus no basis for proposing that it is underpaid. Secondly, in relating the monetary cost of labor to the price of commodities, it is not clear why labor should be singled out among the various costs of production—including, for instance, fuels and raw materials—as the one factor which allows the capitalist to profit from the difference between costs of production and proceeds from sales. To reiterate Marx's well-known narrative about the worker who is only paid for a part of his work day is no more convincing than to say that fuel costs only cover a part of their contribution to the production process. What labor and fuels have in common is that they are both forms of energy employed in production. To suggest that the use of labor energy has a specifiable relation to capitalist profits on the world market is analytically indistinguishable from the suggestion that energy in general has this ubiquitous connection to the augmentation of utility and to income from sales. It is thus not surprising that a leading theorist of Marxian economics, focusing particularly on the relation between Marxist theory and ecology, should now discover a fundamental agreement between labor and energy theories of value (Foster 2013; Foster and Holleman 2014). As Lonergan (1988) observed long ago, Marxist and ecological approaches to unequal exchange—conceptualized as underpayment of labor and energy, respectively—are analytically identical.

Flows of Energy and Flows of Money Signs: Historical Disjunctions and Conceptual Confusions

The confusion regarding the relation between biophysical factors of production such as energy, on the one hand, and monetary, economic growth or capital accumulation, on the other, became particularly pronounced in the merchant capitalist states of early modern Europe. Whereas most societies until then had shared an intuitive acknowledgment of the sun's energy as the vital essence flowing through all living things, the experience of long-distance traders instead suggested that the essential flow was that of money. This certainly became a predominant worldview in the Portuguese, Dutch, and British trading empires, and to this day it no doubt remains a perplexing question for most people whether energy or money is ultimately the most important vital flow animating human society. A reasonable response today would be that the significance of money is precisely that it can provide access to energy, indicating that energy in the final instance is more indispensable than money. In the eighteenth and nineteenth centuries, however, the Physiocrats and Marxists had great difficulties reconciling the physical and semiotic aspects of economic growth. Both agricultural and industrial economies were based on material processes of production requiring physical inputs, yet the market valuation of their produce—and thus their income from sales—hinged on semiotic processes determining people's willingness to pay. The concept of economic value belongs to the vocabulary of the market. The ambition to explain economic value in terms of physical inputs, whether of labor, land, or more generally energy, is to confuse two levels of reality that ought to be kept analytically distinct. The recent debates between Marxists and ecological economists reviewed here illuminate this ancient source of confusion.

Rather than engage in further exegesis, it will suffice to outline the essential differences between the four main positions in these debates (Table 5.1).

Table 5.1 Some Essential Differences Between Four Traditions of Economic Thought

Issues	Neoclassical economics	Marxism	Neo-Physiocrat ecological economics (e.g., H. T. Odum, R. Costanza)	Nonreductionist economics (N. Georgescu-Roegen)
How is economic value defined?	By consumer preferences	By the quantity of embodied labor time	By the quantity of embodied energy or other natural values	By consumer preferences
Why are there environmental problems?	Environmental costs are insufficiently internalized in market prices	The capitalist mode of production generates environmental destruction	Natural values are insufficiently internalized in market prices	Economic value creation generates entropy
What are the prospects of technological progress?	Technological progress will solve all problems	Technological progress and a shift to socialism will solve all problems	Technological progress and restraints on consumption can solve all problems	The prospects of technological progress are limited by the law of entropy
What is unequal exchange?	A result of market power, such as monopoly	A result of the underpayment of labor	A result of the underpayment of natural values	A result of the interaction of market valuation and physical laws

The differences reflect internally coherent frameworks of thought in each of the four schools and are reflected in their distinct approaches to unequal exchange. It will be noted that different foundational assumptions unite different traditions of economic thought. What I refer to as "neo-Physiocrat" ecological economics, which reduces economic value to physics, thus tends to share with neoclassical economics the understanding of environmental problems as the result of insufficiently internalized ecology, whether conceptualized as environmental externalities, ecosystem services, natural values, or embodied energy, and with Marxism a materialist approach to economic value and unequal exchange. Whereas neoclassical economics in general does not consider material constraints on economic processes, Marxism and both varieties of ecological economics aspire to unravel how their monetary and material aspects are related. However, only in the nonreductionist economics pioneered by Georgescu-Roegen do we find a consistent analytical distinction between the semiotics of market valuation and its material consequences. His conclusion, that *the products of economic processes simultaneously represent greater consumer value and greater entropy than the inputs in such processes*, remains a formidable challenge to any advocate of economic growth and technological progress.

In consistence with the different approaches to economic value and environmental problems embraced by the four schools in Table 5.1, each school offers a distinct perspective on unequal exchange. In neoclassical economics, unequal exchange is acknowledged only under conditions of market power, such as monopoly. In Marxist and "neo-Physiocrat" ecological economics, it is viewed as the result of underpayment of labor-power and natural values such as energy, respectively. Although not explicitly stated in the nonreductionist ecological economics of Georgescu-Roegen, his theoretical framework should imply an approach to unequal exchange that views it as *a result of the interaction of market valuation—in general-purpose money—and physical laws*. In articulating the cultural and political determination of commodity exchange values with inexorable processes of material resource degradation, it illustrates how social and natural realities must be kept analytically distinct if we are to grasp how they are intertwined.

To summarize the main points of contention vis-à-vis the heterodox perspectives on economics addressed in this chapter, we may conclude that to refer to asymmetric flows of energy, materials, or embodied land as an unequal exchange of use values confuses physics and economics. The concept of "use value" should refer to what people find useful, or what conventional economics calls utility. It is thus defined by the cultural semiotics of consumption and cannot be measured in biophysical metrics such as Joules, tons, or hectares. Such metrics, however, are the only tools we have to demonstrate the occurrence of ecologically unequal exchange. This is why the Marxian notion of underpaid use values fails to help us establish the material asymmetry of world trade. There simply is no way of deducing any kind of economic value from invested Joules, tons, hectares, or hours of human labor. Rather than aspire to an economics that is congruent with physics, our concern should be to design an economy that radically reduces the physical vulnerability of humans. To sketch the outline of such an economy is the aim of the final chapter in this book.

CHAPTER 6

Agency, Ontology, and Global Magic

In 1909, a Maori native of New Zealand by the name of Tamati Ranaipiri explained to ethnographer Elsdon Best why, when a person receives a gift from another, he is obliged to give something in return. Ranaipiri told Best that the return gift is the spirit (*hau*) of the original gift and that it would be unfair and possibly harmful not to reciprocate. Mauss used this account as point of departure for his classic essay on *The Gift* (1990 [1925]), which for almost a century has been foundational for a vast literature on economic anthropology. The ethnographic and theoretical complexities of these deliberations shall not detain us in this little book, but the volume and intensity of discussion unleashed by Mauss's exposition of the principles of premodern exchange reflect the profound significance of the contrast he established between gifts and modern commodities. Mauss (1990 [1925]: 12) concluded that Maori gifts create ties between human souls "because the thing itself possesses a soul," and "to accept something from somebody is to accept some part of his spiritual essence, of his soul. To retain that thing would be dangerous and mortal, not only because it would be against law and morality, but also because that thing coming from the person not only morally, but physically and spiritually . . . exert[s] a magical or religious hold over you."

As suggested in previous chapters, the attribution to artifacts of such autonomous agency is a fetishized representation of social relations. In order for artifacts to so efficaciously record and buttress the webs of human strategies, the items themselves are accorded significance beyond that of mere objects. As signs of relations, they are identified with those relations, inducing the emotions and moral obligations that are evoked by other humans. The management of such artifactual signs assumes a logic and a purpose of its own that define the social game and rules of proper conduct. Social relations are embodied in artifacts, and the management of artifacts is tantamount to the management of relations. This is a pervasive revelation of economic anthropology, epitomized by the exchange of *kula* valuables in Melanesia. There is a widespread inclination for the preoccupation with material items of exchange to usurp other aspects of interpersonal interaction. The question to be addressed in this chapter is how the transition from premodern to modern economies has transformed these conditions of human sociality, and how this transformation is manifested in modern technology. If modern money, commodities, and technologies are *not* understood to exert a "magical or religious hold" over us, what does it mean to say that we have dismissed this belief in the magical agency of objects?

Modern money and commodities, Marx observed, appear to us disembedded from the social exchanges which they represent. In referring to this perception of money and commodities as "fetishism," he suggested that the detachment of these objects from social relations has a magical aspect. Our engagement with such artifacts indeed mystifies the underlying social relations, as illustrated by the moral and psychological dimensions of debt (Graeber 2011a). The modern preoccupation with money and consumption is easily defamiliarized in terms of its "magical or religious" aspects. Nevertheless, modern artifacts are generally decidedly less charged with agency than premodern Maori gifts. They are rarely believed to possess souls or subjective intentions of any kind, whether benevolent or malicious. To the extent that they are attributed autonomous agency, it is impersonal and incapable of afflicting intentional harm. If modern money, commodities, and technologies do represent a kind of magic, it is a variety of magic that differs in fundamental ways from that of the premodern Maori.

In learning to approach artifacts as morally neutral, nonsubjective tools for social interaction, modern people effected a paradoxical transformation of society. By abandoning the identification of artifacts with the social relations which produced them, they were able to dismiss the kind of magic that moderns tend to classify as "superstition," but precisely in detaching objects from relations they were simultaneously able to morally neutralize them. This epistemological shift unleashed new kinds of social games unfettered by moral concerns, most centrally "the economy" and "technology." In unprecedented ways, the logic of money and the modern market constituted a new framework for managing artifacts and social relations. Divorced from any lingering concerns with reciprocity, and given viable rates of exchange, the myriad of commodities traded on the market could be recombined into new "technologies" which fundamentally reorganized the rules of the game while completely obscuring the exploitative relations which made them possible. The eighteenth-century coal miners and enslaved cotton harvesters of the Industrial Revolution were among the earliest victims of the new world market. The disembedding of the economy and the disembedding of technology were mutually reinforcing processes. In historical retrospect we can conclude that to dismiss a belief in the magical agency of traded objects was tantamount to abandoning concerns with reciprocity. The challenge for a modern concern with sustainability and equity is to recharge our artifacts with moral imperatives while retaining the Enlightenment rejection of their magical agency. Both goals are consonant with a rejection of magic, because the illusion of moral neutrality is generated by the new kind of magic which Marx aptly referred to as fetishism.

The capacity to obviate concerns with reciprocity was derived from the phenomenon of money. The idea of money, initially introduced to facilitate exchange at the margins of culturally integrated societies, promoted the conception and practice of a disembedded economy. It finally became incorporated into the very core of mainstream Western society, but not without centuries of traumatic social transformations and moral contradictions (Polanyi 1957 [1944]; Bloch and Parry 1989). Money made concerns with reciprocity superfluous precisely because it was conceived as a means to guarantee it. A market transaction mediated by money is conducted on the assumption that both parties will be satisfied. The circumstances determining which specific exchange values will be mutually acceptable are quite separate from the substance of the exchange itself, whether gauged, for instance, in embodied energy, land, or labor time. This means that the superficial reciprocity of market exchange can orchestrate systematically asymmetric exchanges of material resources which provide opportunities for the local accumulation of technological infrastructure. The asymmetrically exchanged resource that concerned classical economists such as Smith, Ricardo, and Marx was embodied labor, while the tradition of ecological economics has been concerned with flows of materials, energy, and other products of land. However, all such concerns with the substance of exchange have, for the past two centuries, generally been dismissed by mainstream economics and continue to be classified as "heterodox." Since the apex of British colonialism in the late nineteenth century, the preoccupation of orthodox, neoclassical economics has been the determination of market prices. In thus delegating concerns with reciprocity to the market, modern people have devised a new way of distorting the ancient preoccupation with mutuality. Underneath the superficial symmetry of compromise lurk the calculated asymmetries of exploitation.

Animism, Perspectivism, and the Ontological Turn

It has often been asserted that anthropology should be not only about understanding the life-worlds and mindsets of other people, but ultimately about using such understandings to better grasp the cultural specificity of the *familiar*. George Marcus and Michael Fischer (1986) have called such a U-turn of the anthropological gaze "defamiliarization by cross-cultural juxtaposition."

This chapter attempts to defamiliarize the way in which most modern people approach technological artifacts. To unravel how humans deal with artifacts is to unravel the specifics of social relations. The argument in this chapter is that an analytical distinction should be recognized between two very different ways of delegating agency to artifacts, depending on whether such agency is contingent on subjective human perceptions or merely on the physical properties of the artifacts themselves, as in the case of simple tools. Following Marx's insight that artifacts perceived to have intrinsic or magical agency (i.e., fetishes) are pivotal components of political economy in both premodern and modern economies, the aim is to show that the agency of modern technological objects is *not* intrinsic to those objects—and independent of human perceptions and deliberations—but that our belief that *this is indeed the case* is our way of distinguishing the modern from the nonmodern.

Although this may sound much like what Bruno Latour and other proponents of "the ontological turn" have been saying, the argument here is in fact quite different, grounded as it is in political economy. Rather than take ontological differences as a point of departure, the proposition is that we investigate the political economic conditions that produce particular ontologies. This applies no less to ontological diversity among nonmodern societies—for instance, between indigenous Amazonia and the prehispanic Andes—than it does to differences between the modern and the nonmodern. Political economy fundamentally concerns the social organization of human–object relationships, and thus ultimately how social agency is delegated to artifacts. Such a definition of political economy inevitably implicates our own cultural constructions of technology (cf. Pfaffenberger 1992; Hornborg 2001a). In unraveling the difference between two kinds of artifactual agency—that is, whether or not it is contingent on human subjectivity—we discover that the distinction between "subject" and "object" is much too significant to discard, if we want to understand how relations of social power are embodied in technologies. Paradoxically, although Latour's focus on artifactual agency is supremely valid, his aspiration to abandon subject–object distinctions presents an obstacle to analyzing the historical transformations of such agency. To unravel this paradox, we shall need to discuss differences between some of the main protagonists of the so-called ontological turn in anthropology (Bruno Latour, Philippe Descola, Eduardo Viveiros de Castro, and Eduardo Kohn). Whereas Latour's rejection of subject–object distinctions is contradicted by his fellow ontologists, Descola's structural analysis of ontologies is less concerned with the role of artifacts. These omissions mean that neither Latour's nor Descola's framework can in itself adequately account for historical transformations of political economies and their associated ontologies. Moreover, it will be evident that the ontological turn, although an ambitious attempt to challenge

Agency, Ontology, and Global Magic • 97

the hegemony of mainstream Western science and technology, does not represent a coherent or unitary theoretical framework.

To trace the emergence of the contemporary preoccupation with ontology, we shall begin by going back 20 years to June 1994, when, according to Signe Howell, the organizers of the third meeting of the European Association of Social Anthropologists, in Oslo, were taken by surprise by the unexpected interest in the "outmoded" theme of *ecology*. Two years later, Philippe Descola and Gísli Pálsson (1996) gathered several of the papers presented in Oslo in a volume called *Nature and Society: Anthropological Perspectives*. What most of the papers in the volume had in common was an understanding that the conventional nature–society or nature–culture dichotomy so prominent in European thought can generally not be identified ethnographically among indigenous, nonmodern populations in, for example, Amazonia, Southeast Asia, or Oceania. Some papers, frequently citing Latour, also addressed recent trends toward blurring the nature–culture opposition in contemporary science, prompting the editors to ask whether this will "imply a redefinition of traditional western cosmological and ontological categories" (Descola and Pálsson 1996: 2). No longer simply relegating concerns with a modernist concept of ecology to the margins, constructionist and culturalist approaches in anthropology were now prepared to apply their perspectives to human–environmental relations and to nature itself.

The papers assembled by Descola and Pálsson in 1996 were foundational to the wide-ranging discussions on animism, perspectivism, and human–environmental relations that have preoccupied so many anthropologists since then. Descola's (1996) structural analyses of what he calls "the social objectivation of non-humans [as] a finite group of transformations" have developed into the canonical volume *Beyond Nature and Culture* (2013), characterized in the foreword by Marshall Sahlins (2013) as "a comparative anthropology of ontology" and nothing less than a paradigm shift. Descola's quadripartite typology of ontologies—naturalism, animism, totemism, and analogism—is elegantly generated by the logical intersection of two parameters: here, continuity versus discontinuity in the representations of "interior" versus "physical" aspects of existence. However, Sahlins (2014) has suggested that Descola's categories animism, totemism, and analogism ultimately are merely three varieties of animism, all founded on a general inclination toward anthropomorphism. This conclusion is congenial with the proposal, in this chapter, that Amazonian animism and Andean analogism should be more closely related than Descola's analysis suggests. It also confirms that the crucial ontological distinction is that between animism and naturalism (cf. Descola 2013: 172). While Descola's empirically rich and theoretically sophisticated analysis is a magnificent account of global variation in human conceptualizations of their nonhuman environments, we shall suggest, in very general terms, how it might be complemented with perspectives linking such conceptualizations to political economy.

Århem's (1996) chapter "The Cosmic Food Web," which elucidates the eco-cosmology of the Makuna, was a significant source of inspiration for the model presented by Viveiros de Castro (1998) in his celebrated article on Amerindian perspectivism in the *Journal of the Royal Anthropological Institute*. Judging from the extent to which the perspectivist model has been adopted and endorsed by other anthropologists, it has had an irresistible appeal to our profession. This appeal reflects not only how impressed we are by the elegant cognitive twists of structuralist methods, but perhaps even more so the way the model enlists indigenous cosmologies to challenge the mindsets of capitalist modernity. To find, in the indigenous Other, the diametrical inversion of the civilization that many of us deplore is arguably a hallmark of much anthropology. The perspectivist model continues to haunt us, perhaps because it recognizes the possibility of acknowledging, in general terms, the subjectivity of all living things, which has been so bluntly repressed in modern society (cf. Kohn 2013).[1] It illuminates how Cartesian objectification of human and nonhuman Others is ultimately an act of moral *dissociation* (Hornborg 2014a).

Although closely related to Descola's understanding of animism, perspectivism was contrasted against the latter in a debate in Paris chaired by Latour in January 2009. Latour's (2009) brief review of the debate presents Descola's approach as the more traditional, preoccupied with ordering typological categories in a "cabinet of curiosities," whereas perspectivism to Latour represents a "bomb" aiming to explode the philosophical typologies ultimately deriving from Western colonialism. Even if Descola's (2013) classification of human ways of relating to nature includes scientific objectivism—he calls it naturalism—as merely one of four ontological options, Viveiros de Castro proposes an even more radical departure from the nature–culture dualism of conventional Western science: the complete dissolution of the notion of an objective, universal nature. Instead of assuming that there is only one nature, but many cultures, he argues, indigenous Amazonians hold—and he obviously thinks that we should take this assertion seriously—that there is only one culture (or spirit, or soul) but many natures—many different material, bodily forms united by a single and shared form of subjectivity.

Descola's and Viveiros de Castro's challenges to Western science are enthusiastically endorsed by Latour. Latour's own work has addressed topics generally classified as Science and Technology Studies (STS), but he often presents his influential deliberations on the philosophy and sociology of science as anthropology. Alongside Descola and Viveiros de Castro, Latour personifies what is now being called the ontological turn in anthropology. Although considerable efforts are being made to persuade the anthropological profession that this turn indeed represents a significant shift away from whatever anthropology used to be, it is not altogether easy to grasp what the professed shift is all about (cf. Vigh and Sausdal 2014). Part of the confusion derives from Latour's contradictory notion of "assemblage," which suggests that the very material, nonhuman phenomena whose agency he wants us to acknowledge can only be said to *exist* in terms of how humans perceive them (Elder-Vass 2015).[2] But much of the confusion regarding the ontological turn stems from significant differences between the main protagonists. For instance, whereas Latour completely rejects the subject–object distinction, it has been explicitly fundamental to Viveiros de Castro's (1999) concerns, and implicitly also to Descola's (2013) focus on interiorities versus physicalities. The issue has important implications for our capacity to distinguish between the agency of living organisms and that of abiotic things (cf. Kohn 2013), more specifically between the agency of humans and that of artifacts.[3]

If, as Latour (2005, 2010) has suggested, we are mistaken to think that there is such a thing as society, capitalism, or fetishism, how could his approach help us theorize *power*? The answer should no doubt be sought in his general approach to artifacts. In a paper co-authored with primatologist Shirley Strum in 1984 (Strum and Latour 1987), Latour observes that the key difference between the sociality of baboons and that of humans is that human relations can be anchored to partially independent and fixed points of reference beyond the body, such as language, symbols, and—importantly—material objects. Although it is hardly a new observation that humans distinguish themselves by the extent to which they use language, symbols, and artifacts, Latour's perspective on the agency of artifacts—apparently emerging from his early studies in primatology—encourages us to reconsider the role of specific properties of artifactual assemblages in generating specific varieties of human social organization. An implication of such a stance should be that the power asymmetries addressed in studies of political economy should be possible to trace to specific kinds of human–object relations.

Artifacts, Ontology, and Political Economy in Indigenous South America

If, to a large extent, artifacts—including technologies—are indeed the substance of increasingly complicated human social relations, Latour's preoccupation with their agency within hybrid networks or assemblages is incisive. It raises questions that are central to the ethnography of indigenous Amazonia and the mirror it provides for capitalist modernity. What is the relation between materiality, sociality, and imagination? Or, differently phrased, what is the relation between political economy, magic, and myth? As Santos-Granero (2009b: 19) has implied, the role of artifacts conceived of as powerful agents would no doubt be a key to understanding sociopolitical organization in the more hierarchical societies—or, to use an expression from Stephen Hugh-Jones (2009), more "opulent object regimes"—known to have existed in precolonial Amazonia. Archaeologists speak of prestige-good systems. Prestige-goods such as greenstone amulets (Boomert 1987), shell beads (Gassón 2000), snuff trays (Torres 1987), and feather headdresses (Basso 2011) appear to have been widely circulated in precolonial Amazonia, and their significance for regional social organization, hierarchy, and power should not be underestimated (cf. Hornborg 2005).

The precolonial transformations of Amerindian societies into chiefdoms, states, and empires, such as those encountered by Spaniards in the Andean highlands, hinged on the political economy of prestigious and fetishized artifacts such as the *Spondylus* shells imported from coastal Ecuador (Salomon 1986; Hornborg 2014b). The Thorny Oyster or *Spondylus* generally occurs naturally not much further south than the Gulf of Guayaquil, but was in high demand throughout the Andean area for millennia before the Spanish conquest. Whether in the form of intact shells or fashioned into ornaments, beads, or powder, it has been discovered in a number of archaeological sites ranging from coastal Peru around 2500 BC to Inca-period sacrifices on high peaks in the southern highlands (Paulsen 1974; Pillsbury 1996; Carter 2011). Ethnohistorical sources indicate that *Spondylus* symbolized fertility and water and that one of its primary uses was as offerings to the gods to ensure good harvests (Salomon and Urioste 1991; Blower 2000). Access to items derived from *Spondylus* provided the lords of prehispanic Andean theocracies with a means of claiming prestige and honor in proportion to harvests, and thus to establish claims on the labor of their dependent peasants. The social and political agency of these small but highly valued fetishes was thus formidable. Much like money in our contemporary world, they integrated vast imperial hierarchies ultimately because most people believed in their magic.

The cultural continuities linking Amazonian and Andean societies have intrigued a number of anthropologists working on both sides of the *montaña*, including Lévi-Strauss. The difference between Amazonian animism and Andean analogism identified by Descola (2013) can no doubt be illuminated by focusing on historical transformations in the political economy of human–object relations in the two regions. We may begin by asking what the relationship is between ontology and political economy. The fifteenth-century capacity of *Spondylus* shells to mobilize thousands of Andean peasants was contingent on how they were subjectively perceived; their symbolic agency was thus distinctly different from the technical impact of other Inca artifacts such as the foot-plow or back-strap loom. In making this difference invisible, a dissolution of the subject–object distinction would also conceal the huge potential for intensification and centralization inherent in what we might call ritual or symbolic technologies. The amount of work that can be accomplished with a foot-plow or back-strap loom per unit of time is limited by the energy and skill of the laborer, but the amount of work that can be mobilized by a gift or sacrifice of *Spondylus* is limited only by human credulousness. The prehispanic ontology of *Spondylus* was thus inextricably intertwined with political economy.

Descola (2013: 268–280) uses Nathan Wachtel's ethnography of the Chipaya in highland Bolivia as representative of the analogism he identifies as prevalent throughout the Andes at least since the Inca Empire. Their dual organizations and quadripartitions are repeated at every level of inclusiveness, organizing society and the cosmos as a consistent, fractal hierarchy that pervades both human and nonhuman domains. As Descola (2013: 202) suggests, the obsession with resemblances in such stratified societies is a way of making a world of "infinitely multiplied" differences intelligible and meaningful, but we also need to ask how those differences were generated that needed to be made meaningful. In the current context, this means asking how the social organization of artifacts and human–object relations in the precolonial Andes could generate vast imperial hierarchies among populations who adhered to a fundamentally egalitarian and reciprocal cosmology. The archaeological reconstruction of the emergence of prestige-good systems addresses precisely this issue: how the expanded circulation of subjectivized artifacts generated new and more hierarchical forms of social organization in prehistory. The political economy of fetishized valuables was a crucial foundation of Andean civilizations (Hornborg 2014b). It is reasonable to hypothesize that such human–object relations have emerged from relations similar to those that are currently being investigated in the less hierarchical indigenous societies of contemporary Amazonia (Santos-Granero 2009a). Rather than understand the difference between Amazonian animism and Andean analogism as an essentialized contrast in worldview or ontology, the challenge for anthropology should be to account for the difference in terms of historical transformations of social organization.

Indigenous Andean and Amazonian societies have experienced quite divergent postconquest trajectories: while Andean communities have remained integrated in the large-scale colonial hierarchies that replaced the Inca Empire, Amazonian groups have been more thoroughly victimized by depopulation and societal fragmentation. However, archaeological investigations in various parts of Amazonia indicate that, prior to exposure to European colonialism, the region was home to densely settled and hierarchical polities that were no doubt comparable to those of the Andes (for an overview, see Hornborg 2005). Extensive areas of raised fields, anthropogenic soils, and earthworks testify to the precolonial existence of complex sedentary societies in various parts of the tropical lowlands (Balée and Erickson 2006; Schaan 2012). Although most of the prestige goods that circulated in and between these polities would have been perishable, there are archaeological indications of long-distance trade in items such as greenstone amulets, shell beads, and snuff trays (Boomert 1987; Torres 1987; Gassón 2000). As Santos-Granero (2009b: 19) has implied, the contemporary uses of "subjectivized" artifacts among indigenous groups in Amazonia may represent fragmented echoes of precolonial political economy. The agency of such subjectivized artifacts (or fetishes, in Marxian parlance) was no doubt as significant for ancient Amazonian social organization as *Spondylus* shells were for polities in the prehispanic Andes. If, as Descola (2013) proposes, analogist ontologies have emerged to reconcile the myriad differences of stratified premodern societies, the distinction between Amazonian animism and Andean analogism cannot be a timeless, essentialized one, but a postconquest divergence of societies that once belonged to the same continuum. It reflects a difference in degree of hierarchization, but not a difference in the fundamental character of human–object relations.

In considering what the ontological turn might have to contribute to our understanding of such historical transformations, we are struck by two conspicuous omissions in the respective frameworks of Latour and Descola. Latour rejects a distinction between the agency of living subjects and that of abiotic artifacts (cf. Kohn 2013: 91–92), and he would thus no doubt also reject a distinction between forms of artifactual agency based on whether or not they are contingent on human subjectivity. Descola, on the other hand, appears to accept subject–object distinctions, but demonstrates little concern for the role of artifacts and human–object relations in generating different ontologies. In offering an alternative perspective on the political economy of globalized technologies, I shall selectively retain Latour's observations on the pivotal role of artifacts in human social organization and Descola's acknowledgment of subject–object distinctions.

If Sahlins (2014) is correct in suggesting that the essential distinction between divergent ontologies can be reduced to that between naturalism and animism, it is significant that naturalism can be seen as closely related to the emergence of new forms of fetishism that were fundamental to the Industrial Revolution in eighteenth-century Europe. Although naturalism has been represented as a transcendence of premodern, local magic, its approach to the agency of technological artifacts is associated with a different, globalized form of magic. Its ontological foundation is the abandonment of relationism that

I discussed in chapter 1, illustrated by the assumption that objects such as organisms and machines can be fully understood through analysis of their bounded, material forms, detached from the relations that generate them.

Keys and Coins: Technology, Magic, and the Significance of Human Subjectivity

The distinction between premodern and modern hierarchical societies hinges on the different roles of human subjectivity in the kinds of human–object relations that characterize the two contexts. If material objects are mobilized as agents in systems of socioecological relations, we should reflect on the difference between their capacity to operate *without* the mediation of subjective human perceptions, on the one hand, and their capacity to operate *by means of* such mediation, on the other. This difference is fundamental to the way we conventionally distinguish between technology and magic.[4]

With this perspective as our point of departure, it is possible to show that technology is our own version of magic. In this sense, Latour (1993) is right in that modernity is not a decisive break with premodern ontologies. The Enlightenment demystification of premodern magic and superstition was not a final purge of reliable knowledge, but a provisional and politically positioned one. Its understanding of the nature of economic growth and technological progress has been a successful instrument of predatory expansion for core regions of the world-system for over three centuries, but the multiple crises currently faced by global society are an indication of the approaching bankruptcy of this worldview. The components of this failing ontology that seem most imminently in line for collapse are its understandings of money and technology—two kinds of fetishized artifacts widely imagined to have autonomous agency.

How could the ethnography of native Amazonia help us to expose and transcend modernist illusions? To understand Amazonian ontologies in terms of how artifacts are incorporated into the social organization of subject–object transformations can shed light on the specific way in which modern people tend to perceive the agency of their technology. Descola (2013: 405) concludes his book *Beyond Nature and Culture* with the assertion that "it would be mistaken to think that the Indians of Amazonia, the Australian Aboriginals, or the monks of Tibet can bring us a deeper wisdom for the present time than the shaky naturalism of late modernity." It is true that their ontologies cannot be transferred and applied to the predicaments of modern life, but a familiarity with the different ways in which humans can relate to material artifacts increases our capacity to critically scrutinize our own constructions of technology. In fact, cross-cultural variation in the way humans relate to artifacts could probably also be analyzed structurally and typologically as a finite group of transformations, much as Descola has done regarding what he calls the social objectivation of nonhuman nature. One parameter to investigate might be various modes of understanding the relations between objects and the metabolic flows which generate them, as in the distinction between fetishism and relationism. Other parameters might include modes of understanding artifactual agency: whether it requires human delegation, whether it presupposes human beliefs, and whether it implies personhood and intentionality or merely posits soulless causation.

As we have seen, a tenacious illusion of Enlightenment thought is that a boundary can be drawn between material forms and the external relations which generate them, and that the former have an existence that is ontologically independent of the latter. This kind of distinction—the reification of *things*—is more problematic than the distinction between natural and societal aspects. It is the essence of fetishism. Modern people generally perceive tangible objects as given, and as separate from the invisible networks of relations in which they are embedded. Such distinctions alienate humans from nonhuman nature as well as from the products of their labor, because both are perceived as categories of autonomous objects rather than as manifestations of relations. What our technological fetishism obscures from view is that it is as misleading to imagine machines as independent of global price relations and resource flows as it is to imagine organisms as independent of their environments. A tractor without diesel is as inanimate as an organism that has starved to death.

How should we understand the role of human perceptions in granting agency to objects? To illustrate the second set of parameters I mentioned, it will suffice to acknowledge that both *keys* and *coins* have been delegated agency, but of different kinds. Such little pieces of metal can be crucial in providing access to resources, whether by physically opening doors or by social persuasion. The way these metal objects are shaped—whether as keys or coins—have for centuries determined whether they operate as technology or through magic.[5] Coins and keys illustrate how social relations of power in different ways are delegated to material artifacts. They exemplify how such delegation can either be dependent on, or independent of, subjective human perceptions. They thus make very tangible the distinction that Keynes long ago made between "organic" and "atomic" propositions, the truth of the former depending on the beliefs of agents, whereas the truth of the latter is independent of any such beliefs (Marglin 1990: 15). It is with this distinction in mind that we shall now compare the operation of modern technologies with native Amazonian uses of artifacts, as elucidated in Fernando Santos-Granero's (2009a) edited volume on *The Occult Life of Things*.

Objectification and Subjectification: Human–Object Relations in Amazonia

Since long before Latour (1993) launched the notion of a symmetric anthropology, the ethnography of indigenous Amazonia has in many ways provided capitalist modernity with a mirror in which to discover its own idiosyncrasies and blind spots. Here we shall consider what it can teach us about our relations to *things*. As Marx realized through his own brand of symmetric anthropology in the mid-nineteenth century, human relations to things are always about relations to other humans. Applying a concept originally employed by Portuguese merchants to describe the "primitive" religious practices of West Africans, he referred to this as fetishism. Based on such a definition of magic—as the attribution of autonomous agency to artifacts, obscuring the role of human perceptions and strategies—I shall argue that modern, globalized technologies qualify as an example of this phenomenon. This will become clearer as we consider various forms of human–object relations among non-modern populations in Amazonia.

Hugh-Jones (2009) observes that although native Amazonia has generally been characterized as "object poor," there is considerable variation in time and space, and different peoples have quite different "object regimes." For example, the Barasana recognize an important category of valuables that signal group identity and social rank, and the northwest Amazon as a whole, like the upper Xingú, is known for the intense circulation of ritual objects and ordinary possessions. Joana Miller (2009: 76) similarly observes that, in regions where they are involved in wider trade and exchange networks, objects produce distinctions within or between social groups, and Terry Turner (2009: 162) shows how traditional Kayapó valuables are passed on over the generations as tokens of social identity. These observations contrast, for example, with those of Philippe Erikson (2009: 177) among the Panoan-speaking Matis, for whom "all artifacts are conceived of as an extension of their maker, and as such, as 'inalienable' extensions of their person." As several of the chapters suggest, the extent to which possessions are alienable from their owner can be expected to be reflected, in related ways, in separate contexts such as the propensity to exchange them and their disposal in connection with funerals.[6]

Erikson (2009) suggests, like Luiz Costa and Carlos Fausto (2010), that the widespread Amazonian concern with mastership or control over humans, animals, and plants also applies to artifacts. This Amerindian "idiom of power" evoking master-ownership, engendering, and protection organizes relations between hunters and their game, warriors and their enemies, chiefs and their followers, shamans and spirits, humans and pets, and parents and children, as well as between persons and things. It is in fact also evident in historical accounts of the relation between the Inca emperor and his *yanacona* servants and in the relation between captors and captives in Santos-Granero's (2009c) survey of precolonial Amerindian slavery. A central component of this pervasive notion of mastership is the capacity to *dispose* of persons and things. Much as when David Graeber (2011b) argues that the sovereignty of the modern consumer over his or her commodified objects is modeled on the sovereignty of the medieval monarch over his or her subjects—as both represent an urge to destroy in order to gain recognition and identity—we find once again that human relations to *things* are about relations to other human or nonhuman beings. Graeber's reflections on the modern concept of consumption as based on the metaphor of eating—the perfect idiom for destroying something while literally incorporating it—are strangely familiar to Amazonianists routinely discussing native concepts of predation as incorporation. In both modern and native Amazonian cosmologies, it seems, incorporation is fundamental to identity.

Another reflection stimulated by these illustrations of how power over objects is in fact power over other subjects concerns the very widespread Amerindian myth about "the revolt of the objects" (Santos-Granero 2009a: 3). Jeffrey Quilter's (1990) article in the first issue of *Latin American Antiquity* on the identification of this myth in ceramic iconography from the ancient Moche culture (AD 200–700) on the north coast of Peru, reflecting on the detailed depictions of animated artifacts battling with their human makers, evokes associations to Hollywood productions like *Terminator*. The attribution of agency and subjectivity to artifacts obviously has the potential to rouse fears that the objects will assume power over their makers. Common to the mythological revolt of the objects and *Terminator* is the fear of an inversion of the social relations of power. The latter case clearly seems to reflect the highly ambivalent fascination with technology on which capitalist modernity is built, but the way it deals with subject–object transformations can be viewed in a new light when illuminated by the social life of artifacts in native South America. Moche iconography from the middle of the first millennium AD clearly illustrates that subject–object distinctions were far from insignificant for precolonial Amerindians.

The main theme running through *The Occult Life of Things* is how objects are attributed with subjectivity. The concepts subject and object are highly contested modern categories, but any attempt at cross-cultural comparison will require an explicit baseline of such fundamental categories through which particular life-worlds can be compared. Without the cognate terms interior versus physical as baseline, for instance, Descola's (2013) comparative analyses would have been impossible. It is one thing to observe the psychological, social, and indeed quite material consequences of *perceiving* certain objects as subjects, and another to account for such *perceptions* in terms of the observer's own assumptions about what subjects and objects actually are.[7] Beyond human perceptions, it is undeniable that there are objectively biotic versus abiotic entities, and any attribution of agency or personhood to abiotic objects—whether by Amazonians or by STS scholars—should be understood as a statement about fetishized social relations.[8] The attribution of subjectivity to objects (or vice versa) is a powerful and very real aspect of social causality, but for an observer to acknowledge the repercussions of such attributions is not necessarily equivalent to abandoning his or her own categories of subject and object.

Viewed from within a particular human life-world, objects can be turned into subjects, and vice versa. Rather than discussing the conditions of subjects and objects as nouns, it is thus apt to consider them as *verbs*—as processes of subjectivation and objectivation that must be continuously attended to, through myriad practices including shamanism, ritual, dieting, and daily routines. Such recurrent practices involve acts of subjectivation as well as desubjectivation, exemplified by the neutralization of potentially dangerous food, or the destruction of a dead person's possessions. As Harry Walker (2009) shows, even where potentially dangerous objects are allowed to maintain a measure of agency, they need to be tamed or "subjected."[9]

Much as Viveiros de Castro's (1999) seminal analysis of indigenous Amazonian ontology, Santos-Granero's (2009a) collection of perceptive ethnographies is couched in the inescapable, naturalist language of subjects and objects. The animist perception of all living things as subjects is perfectly compatible with the perspective of *ecosemiotics* (Hornborg 1996, 2001b; Kohn 2013), but the attribution of subjective agency to abiotic artifacts is more correctly classified as fetishism (cf. Gregory 2014). Whether we are confronted with the nonmodern subjectivation of objects or the claims of Actor-Network theorists, we need to retain the capacity to distinguish between sentient actors pursuing their purposes, on the one hand, and objects that simply have consequences, on the other. Kohn (2013: 91–92) thus pertinently criticizes STS for not distinguishing between the agency of sentient selves versus the mere material resistance of abiotic things such as rocks or artifacts. The assertions of ANT about the agency of artifacts, combined with its dismissal of subject–object distinctions, are tantamount to fetishism (cf. Gregory 2014; Hornborg 2014a; Martin 2014).

It is symptomatic of the ontological turn, however, that no one any longer seems to want to talk about fetishism (Goldman 2009; Latour 2010). The implicit assumption is that if objects are perceived as subjects, then *who are we* to suggest that it is an illusion? We are *all* fetishists, says Latour (2010). Yes, Marx said the same thing 150 years ago, but the crucial difference is that he wanted to *expose* fetishism in order to ultimately reject it. He observed that fetishism—the attribution of properties of living things to inanimate objects—could be a means of maintaining social relations of power and inequality. This does not seem to concern most Actor-Network theorists, whose arguments instead tend to amount to an *endorsement* of fetishism. It may be politically correct not to impute fetishism to others, but what is the bottom line of this argument if it simultaneously means denying us the chance of exposing our *own* fetishism?[10] We must reject the approach, expressed, for instance, by Goldman (2009), of denying anthropology any other function than to communicate (and endorse?) non-Western ontologies.

In his introductory chapter to *The Occult Life of Things*, Santos-Granero (2009b) sorts out the various ways and contexts in which objects are attributed with subjectivity in Amazonian societies. He notes that some objects are conceived of as persons because they are attributed with a soul and a measure of agency, but agrees with María Guzmán-Gallegos (2009) that objects can be perceived as subjects even when they are not believed to have a soul, and that their agency does not necessarily imply intentionality. As we shall see, it is significant that these qualifications about soulless subjectivity and nonintentional agency appear in a context where indigenous people are being engaged in the operation of capitalist modernity, quite reminiscent of Michael Taussig's (1980) account of the baptism of money.[11]

A recurrent phenomenon in native Amazonia is the notion that the subjectivity of objects is an extension of the people who made them, which of course recalls Mauss's (1990 [1925]) classic observations on the spirit of the gift. As illuminated by Joanna Overing, Cecilia McCallum, Els Lagrou, and other ethnographers of Amazonia, artifacts and children are often viewed as analogous fabrications, both embodying the extended subjectivity of their makers. But objects can also become gradually ensouled through contact with their owners, whether or not the owner was also the manufacturer.[12] Some objects need the intervention of humans to activate their agency. The only objects that are recognized as completely inanimate are those with which no communication is possible. As Santos-Granero (2009b: 11) observes, "Some objects are just plain objects."

Technology as Magic: Fetishism in Capitalist Modernity

Whereas modern people would generally consider the treatment of objects as personified subjects an illusion or fallacy likely to be dismissed as superstition, while perhaps conversely challenging the objectification of subjects—such as animals or workers—as indicating a lack of empathy, native Amazonians take seriously the risks inherent in such subject–object transformations.[13] They are, in short, concerned with managing *relations*. Such a relational epistemology, as Nurit Bird-David (1999) has called it, is indeed very different from the rigid subject–object dualism which is so diagnostic of modernity.

A fundamental paradox of capitalist modernity, which we can detect in this cross-cultural mirror, is that its naturalist categories of subject and object are so *irrelevant* to the systems of relations that it organizes—that is, in terms of how subjects are *treated* and objects *understood*. Not only does it objectify both human and nonhuman subjects, and treat humans and nature accordingly, it is equally founded on an unprecedented subjectification of objects. This is not to suggest that objects are generally attributed with personhood, but some objects are attributed with an autonomous agency, which serves to mystify unequal social relations of exchange. As Marx observed, money is thus believed to generate more of its own kind, when deposited in bank accounts. Machines are believed to work or produce on their own account, regardless of the global price relations which make them possible, and which should prompt us to understand them as accumulations of embodied human labor and natural resources *where the money is*. Money and machines may not be ensouled *persons* in modernity, but they are certainly believed to have autonomous agency. We pride ourselves on having abandoned animism, but have organized a global society founded on fetishism. It is a fetishism which differs from pre- and nonmodern forms of fetishism by restricting the subjectification of objects to imputing agency to them, rather than full personhood and intentionality, but it is fetishism all the same.

We are now in a position to draw more precise conclusions on the difference between capitalist modernity and native Amazonia in terms of how humans tend to subjectivize artifacts. The distinction between magic and technology that I have suggested corresponds to a distinction between societies founded on the energy of human labor, on the one hand, and societies founded on the use of *exosomatic* energy (primarily fossil fuels), on the other. Where political economy is about the social organization of human muscle power, people have to be *persuaded* to exert themselves for the benefit of those in power. Magic could be defined as the category of social strategies by which such persuasion is achieved.[14] For example, when the Inca emperor offered Ecuadorian *Spondylus* shell to the gods to ensure rain and agricultural fertility, it was incumbent on his many subjects to labor on his terraces and irrigation canals. We can now conclude that the efficacy of such ritual sacrifices was dependent on human perceptions. The prehispanic agency of *Spondylus*, like that of modern money, was contingent on human subjectivity. But when modern farmers in an increasingly desiccated California resort to high-power water pumps to irrigate their fields, the efficacy of such practices is *not* perceived as dependent on human perceptions. The difference between magic and technology, we tend to believe, is that the latter is a matter of increasingly sophisticated inventions based on discoveries about nonsocial nature, which grant our economies the capacity to grow on their own account.

But neither did the peasants of sixteenth-century Peru believe that the efficacy of ritual sacrifices was dependent on human perceptions. The efficacy of all magic hinges on it being *perceived* as independent of human consciousness. Like magic, power over other people is universally mediated by human perceptions (cf. Graeber 2001: 245–246), but this is *never conceded*, except in retrospect. Would it be possible to argue that modernists are as deluded by the magic of their artifacts as any premodern people ever were? Can we manage to expose the magic of our technology? Fundamental to such a shift of perspective are the implications of realizing that global price relations are systematically excluded from our definition of technology, even though, by organizing asymmetric resource flows, they are crucial for its very existence. Without a doubt, Cartesian dualism is at the root of the difficulties we have in perceiving our technological fetishism.

When the Inca emperor imported *Spondylus* shells from Ecuador to persuade his subjects to labor in his fields, the productive potential of *Spondylus* was symbolic—it was dependent on human perceptions. When the California farmer imports oil to run his water pumps, the productive potential of oil appears to be objective, like turning a key in a lock, independent of perceptions. But here is the illusion of modern technology: his *access* to oil, and to the machinery it animates, is ultimately contingent on the socially constructed rates by which oil is exchanged for American exports on the world market. And whatever economists will tell us, we should never doubt that those rates are dependent on human perceptions.[15] Locally, our technology mystifies us by pretending to be productive independently of exchange rates, but viewed from a global perspective, it is indeed dependent on human perceptions.

A conclusion from these deliberations is that we should distinguish between three fundamental categories of artifacts, defined by the specific ways in which they are delegated agency. The first is local, nonindustrial technology, which operates without the mediation of either human perceptions or exchange rates. It can be exemplified by keys or by locally produced implements such as the Andean foot-plow. The second is local magic, which operates by means of human perceptions, exemplified by coins or *Spondylus* shells. The third is globalized technology, which *locally* appears to operate without the mediation of human perceptions, but globally relies on exchange rates continuously shaped by the strategies of market actors.[16] It could also be called global magic, and can be exemplified by machines such as water pumps that run on fossil fuels or electricity. If we do not retain our capacity to distinguish between the subjective and the objective, the crucial differences between these three categories of artifacts will remain invisible for us.

We must conclude that, from a global perspective, modern technology *is* magic. It is a specific way of exerting power over other people while concealing the extent to which this is mediated by human perceptions. In addition to sketching this argument for a radical revision of our Western worldview, we have found that some tenets of the so-called ontological turn in anthropology are not necessarily very helpful in constructing such an argument. Ultimately, the confrontation between Amazonian animism and Euro-American naturalism is a political issue, where the claim of modern science and technology to be objectively superior has proven difficult to deconstruct. Unfortunately, appeals to the virtues of animism are not likely to turn the tables on capitalism. But if Amazonian anthropology can provide us with the analytical tools to demonstrate that the Euro-American technology which is now devastating the Amazon Basin is itself a kind of magic, it would be an irony that I think many anthropologists—and many Amazonians—would appreciate.

More ominously, this conclusion suggests that the pervasive assumption of technological progress as the salvation of industrial civilization is no less naïve than other cultural illusions that have sustained premodern empires facing collapse. As our anxieties about the future prospects of this civilization become increasingly difficult to suppress, there emerges the contrary, neoromantic sentiment that indigenous, animist ontologies could provide us with clues on how to achieve sustainability and resilience. But rather than championing a magical ontology that most of us have irrevocably lost, an anthropological approach is more usefully applied to exposing the unacknowledged magic of our own ontology. Although the project of defamiliarizing and deconstructing our presumptively modernist categories is very much facilitated by juxtaposition with nonmodern ontologies, this is not necessarily tantamount to advocacy of the nonmodern, but may well amount to an acknowledgment that our categories have not been modern enough.[17]

CHAPTER 7

The Political Ecology of Technological Utopianism

In 2010, the Bank of America opened a 55-story skyscraper in Manhattan Island, New York, which in the press was praised as "the most sustainable in the country" and as one of the "most environmentally responsible high-rise office buildings" in the world (Roudman 2013). The building had been given a Platinum certification by the so-called Leadership in Energy and Environmental Design (LEED) and was applauded by Al Gore as a model for combating climate change. However, according to an assessment by New York City in 2012, the same building "produces more greenhouse gases and uses more energy per square foot than any comparably sized office building in Manhattan" and "uses more than twice as much energy per square foot as the 80-year-old Empire State Building" (ibid.). The main function of certification schemes like LEED, the journalist Sam Roudman concludes, is to create a market for sustainability and green publicity, rather than to save energy.

In this chapter we shall consider how globalized technological systems, like the economic systems which make them possible, tend to promote social opacity. From the vantage point of individual participants, the aggregate consequences of technologies are virtually impossible to assess. Technologies designed to solve specific problems are routinely revealed to generate other problems, often for other categories of people (cf. McNeill 2000). The global implications of a particular technology can rarely be predicted at the local level where it is designed and applied. The paradigmatic example is the turn to steam power and fossil fuels in early industrial Britain, which was a local strategy for increasing profits, the long-term global implications of which could not be anticipated by individual factory owners (Malm 2016). In this book we have reviewed several of these implications, including fundamental transformations of global political economy and of the mainstream European worldview, but currently the most conspicuous consequences of the Industrial Revolution are its long-term contributions to climate change.

Technological Progress as a Cultural Category, Situated in Global Social Space

Ever since the Industrial Revolution saved Britain from ecological crisis in the early nineteenth century, visions of miraculous new technologies have alleviated Euro-American anxieties about the impending doom of the fossil-fueled capitalism that it inaugurated. Although Malthus's worries about land shortages were transcended by historical events as well as by Ricardo's and Marx's different versions of technological optimism, they were soon reincarnated in Jevons's warnings about the depletion of coal. Today economists generally dismiss the pessimism not only of Malthus and Jevons, but also of current concerns over peak oil, by expressing faith in human ingenuity, whether in the form of solar panels in the Sahara desert or other forms of putatively "green" production such as biofuels. To retrospectively ridicule pessimists by referring to technological progress that they did not anticipate has become an established pattern of mainstream thought. Almost regardless of ideological persuasion, the seemingly self-evident concept of technological progress inherited from early industrialism has been resorted to as an article of faith serving to dispel the specter of truncated growth. The increasingly acknowledged threats of peak oil and global warming are thus generally countered with visions of a future civilization based on solar power. Considering the serious doubts that have been raised regarding the feasibility of solar power as a global solution to future energy crises, it is valid to ask whether this technological scenario should in fact be viewed as unrealistic. The technological utopianism professed, for instance, by some Marxists (e.g., Schwartzman 1996, 2008) raises fundamental analytical questions about the relation between thermodynamics and economics.

Deliberations about technological futures tend to be founded on considerations of what is feasible to achieve, given current or anticipated knowledge. A common proposition is that a given technical process that has been successfully implemented under laboratory conditions, while still incapable of competing economically with conventional technologies, can soon be expected to be economically viable. Such proposals tend to unite engineers and economists under a common paradigm regarding the nature of technological innovation, even if neither profession is actually prompted to consider technological systems holistically, as simultaneously material *and* social strategies. To understand the conditions of technological progress in such a truly transdisciplinary way, we need to raise a very diverse set of questions, ranging from thermodynamics and material resource requirements to financial politics and the global distribution of purchasing power. No single business or research specialization is equipped to articulate an understanding of technological progress that takes such diverse factors into serious consideration simultaneously.

A successful technical experiment does not provide sufficient evidence that a new technological system is "feasible" or "within reach." Yet, it is noteworthy that such conclusions are very frequently drawn in both academic and public debate. But if we agree that technical and societal feasibility are not synonymous—that technical ingenuity is a necessary but not *sufficient* condition for adoption—we need to ask what kind of obstacles might obstruct the emergence and expansion of a new technology, once its purely technical feasibility has been proven? On the one hand, there may be material constraints such as unreliability, natural limits on resource availability, or locally perceived inefficiencies in energy conversion. On the other hand, there may be various kinds of social constraints. First, there may be cultural constraints such as conservatism or the relative aesthetic virtues of competing designs. Second, there may be economic constraints such as high costs, low profitability, and lack of competitiveness. Third, there may be political constraints deriving from ethical considerations, legislation, policy, or trade restrictions. More generally, we must consider how various such social constraints may simply be expressions of the fact that the expansion of a given technology is ubiquitously limited to that fraction of the world's population which has sufficient purchasing power to adopt it. In other words, modern technology is always and everywhere a matter of uneven distribution in global society. This means that the extent to which a given technology is adopted hinges on the distribution of money in the world-system, and that the technology itself represents an unequal exchange of resources between different economic segments of global society.

The conventional scientific and popular understanding of technological innovation is that it increases efficiency in a cumulative development that progresses over time. In the well-known IPAT equation (Ehrlich and Holdren 1971), for instance, technology (T) is assumed to mitigate the environmental impacts (I) of growing population (P) and affluence (A). Counter to this understanding are glaring inefficiencies and unsustainable practices that paradoxically also seem to increase over time, such as waste of resources, environmental degradation, and economic inequalities. These inefficiencies are often referred to as externalities, which might be mitigated by modifying prices. On the other hand, we have suggested that the very rationale of capitalism is to *keep* such externalities external. It has been argued, for instance, that growth-based "dematerialization" and the so-called environmental Kuznets curve is a local illusion, ignoring the displacement of growing environmental loads to world-system sectors with less purchasing power (cf. Fischer-Kowalski and Amann 2001).

Does technological development generally increase efficiency, or does it increase *in*efficiencies? In order to address this issue, two questions should be posed: (1) By which parameters is efficiency defined? Whereas efficiency is generally assessed in terms of inputs and output of exchange values, or money, there is a widespread neglect of other resource metrics such as embodied/ expended energy, materials, human time, and natural space, and of the impacts of production and transports on, for instance, biodiversity, environmental quality, or human health. (2) How are the boundaries defined for the social units assessed? Whereas efficiency may appear to be increasing within a given social unit A, it may be decreasing within a wider social system of which A is a subsystem.

Increased technological efficiency may thus be largely illusory due to (1) an inadequate consideration of all parameters and (2) an inadequate definition of the boundaries of the social unit under consideration. A case study chosen to empirically illustrate such conditions is the adoption of steam technology in British textile production in the nineteenth century (Hornborg 2006[1]). The argument is founded on (1) a consideration of international transfers of embodied human labor time and embodied natural space, rather than exchange value/money and (2) the total implications of this technology within a global system of nations engaged in trade, rather than only within Great Britain. A conclusion of this case study is that it is valid to propose a thorough rethinking of technology as a global social phenomenon and cultural category. Rather than a product of local or national innovation generating an increase in overall efficiency, a global perspective on technological development reveals that, to a considerable extent, it may represent an increasingly unequal redistribution of resources among different sectors of world society. To argue that technological progress in this sense is inextricably connected to unequal exchange requires a fundamental reconceptualization of the relation between physics and economics, even in schools of economic thought that are currently perceived as challenges to mainstream views.

All human societies have run principally on solar-derived energy, whether in the form of food, animal fodder, wind, water power, or fossil fuels. While alleviating some premodern problems, such as the constraints posed by a limited land area for the production of food as well as energy, fossil fuels introduced some new ones, primarily the fact that supplies are limited and the threat of climate change. Although worries about such drawbacks have been voiced from the start, concerns over peak oil and global warming have become significant ingredients of public consciousness in recent decades. Paradoxically, the celebration of technological progress inaugurated with the adoption of fossil fuels—the so-called Industrial Revolution—later provided the predominant template for envisioning their abandonment. Thus, hydroelectric and nuclear power facilities have been advocated precisely because they represent technologically superior and cleaner alternatives to fossil energy, even if most of the infrastructure for such energy sources continues to be built and maintained with the use of fossil fuels. Hydroelectric and nuclear power facilities have yet other drawbacks, which need not detain us here, but which have contributed to the popularity of visions of a future world powered by direct use of solar energy.

Environmentally Benign Technologies as Illusions: The Example of Solar Power

The trust that technological progress will bring us a solar-powered future has been a central strain in much of the debate about global sustainability for several decades. Already in 1902 the president of the American Chemical Society predicted that the United States in the 1970s would run primarily on solar energy (Nader 1996: 262). In the 1950s, the president of Harvard University, James D. Conant, believed that solar power would be the dominant source of energy by the end of the century (Nader 2004: 801). The first photovoltaic cells were mounted in 1958 on an American satellite (Zehner 2012: 17). Over 50 years ago, the cover of Farrington Daniels's book *Direct Use of the Sun's Energy* (1964) proclaimed that the "most plentiful and cheapest energy is ours for the taking." Already at that time, Daniels referred to steady progress in the direct use of the sun's energy during the preceding decade. He asserted that "technologically it could be used to replace the energy now being supplied by fuels and electricity" and predicted that, given more expensive fossil fuels and future development of solar equipment, it would eventually be able to compete economically with fossil fuels (ibid.: 253). Daniels concluded that the main limitations to use of the sun's energy are economical rather than technological, but predicted that markets will materialize first among "developing countries where there are difficulties in international payments" (ibid.: 259). Twenty-three years later, the so-called Brundtland report, *Our Common Future*, predicted that, with "constantly improving solar thermal and solar electric technologies, it is likely that their contribution will increase substantially" (World Commission on Environment and Development 1987: 193). The report asserts that renewable energy systems "offer the world potentially huge primary energy sources, sustainable in perpetuity and available in one form or another to every nation on Earth" (ibid.: 192).

Some years ago, however, a prominent energy expert observed that direct conversion of solar radiation to electricity by photovoltaics "has succeeded only in small niche markets that can tolerate the high cost" (Smil 2006: 188). The two main reasons for the limited commercial success of photovoltaics are low conversion efficiency—the best field efficiencies amounted to less than 10 percent—and the high cost, which still made it unable to compete with fossil fuels. According to one calculation, a global shift to photovoltaic power is estimated to cost around a hundred times the gross domestic product (GDP) of the United States (Zehner 2012: 9). Nevertheless, Vaclav Smil (2006: 203–204) affirms that photovoltaic conversion remains "the most appealing of all renewable sources." Another study concedes that solar power is still relatively expensive today, but predicts that it will be cost competitive by as early as 2020 (Delucchi and Jacobson 2011: 1174).

Since high costs, particularly investments in infrastructure, imply expenditures not only of money but also of the energy it represents, the costs can serve as a point of departure for estimating the *net energy* of an energy source, that is, how much energy is retrieved in relation to the amount of energy invested. Prieto and Hall (2013) estimate the net output of photovoltaic energy in Spain as around 2.4 times the input, but they add that more detailed calculations might even give a figure less than 1.[2] As approximately 86 percent of global commercial energy use is currently (the year 2011) derived from fossil fuels, Prieto and Hall conclude that 86 percent of the sums of money invested in infrastructures for nonfossil energy such as nuclear or solar power represent subsidies from fossil fuels.[3] This means that so-called alternative energy sources should *not* be conceived of as carbon free (Andersen 2013; Prieto and Hall 2013: 118), and that rising oil prices should *not* be expected to make them more competitive. It has recently also been demonstrated that the modest expansion of renewable energy use tends to "simply be added to the energy mix without displacing fossil fuels" (York 2012: 2). The same author acknowledges that photovoltaic power and wind power "require large amounts of material, some of it toxic and energy-intensive to produce, as well as large areas of land to produce substantial amounts of energy" (ibid.: 3; reference to Smil). As David MacKay (2012: 4) warns us, "someone who wants to live on renewable energy, but expects the infrastructure associated with that renewable *not* to be large or intrusive, is deluding himself."

The dismal observations on the modest success of solar power have never deterred grandiose visions. Yet around 86 percent of the world's total commercial energy use remains based on fossil fuels, while only one-thousandth of it is photovoltaic. Of this very tiny fraction, 80 percent is produced in five of the world's most affluent countries (Germany, Italy, Japan, the United States, and Spain), and in none of these countries solar power in 2011 accounted for more than 1 percent of total energy production (Prieto and Hall 2013: 7–9). Even if the relative expansion of solar power may seem rapid, in absolute figures it is very much slower than the global growth of energy demand. In spite of the expectations, solar power continues to be more expensive than the fossil energy which subsidizes it, which means that it is doubtful whether it will be able to replace oil, coal, and gas, and that it continues to be accessible only to the most affluent sectors of the world-system. It illustrates how expensive, environmentally benign technology automatically implicates perspectives from global political ecology and the discourse on environmental justice.

The optimistic forecasts regarding solar energy ubiquitously rely on the promises of anticipated, but as yet unrealized, technological progress. The boundary between technological and social phenomena in these deliberations tends to be blurred and unexplored. Thus, a proposal published a few years ago in *Energy Policy* suggests that the barriers to providing all global energy from wind, water, and sunlight are "primarily social and political, not technological or even economic" (Delucchi and Jacobson 2011: 1170). For many social scientists, however, it is impossible to extricate a purely technological consideration from social, political, or economic ones. The objective materiality of technological objects renders them seemingly autonomous vis-à-vis societal power structures, symbolic systems, and global resource flows, but this autonomy is an illusion. Discussions of technological options that do not consider social power, semiotics, and international trade are thus fundamentally flawed. Jacobson and Delucchi (2011: 1155, 1157, 1159) claim that it is technically feasible to provide all global energy from wind, water, and sunlight, that wind and solar power are "available today," and that solar energy can power the world "15–20 times over," claims which suggest to many social scientists that they and engineers live in different universes. This becomes even more evident when we turn to popular sources such as Wikipedia, where the entry on "Solar energy" claims that its uses are "limited only by human ingenuity," and the entry on "Solar power"—observing that photovoltaic electricity costs four times more than that generated by coal—naïvely notes that "developing countries in particular may not have the funds to build solar power plants" (both accessed on September 2, 2011).

Rather than suggesting a replacement for fossil fuels, solar energy is an expression of the global processes of capital accumulation which fossil fuels have made possible. Moreover, the production of photovoltaic infrastructure generates significant environmental pollution and emissions of greenhouse gases (Zehner 2012: 18–19; Andersen 2013). But to most environmentally engaged researchers, activists, and people in general, solar power continues to represent a sustainable and democratic way of replacing fossil fuels. This illustrates how technology is conceived by most people as based exclusively on revealing the veiled inner essence of nature. The challenge, in this view, is to discover how to get nature to serve us in the way we imagine. Questions whether it is even *possible*—or what the high costs *mean* from a social-science perspective—are not posed. The economic conditions are presented as features of specific technological systems, rather than as expressions of price and exchange relations in global society. The high costs of solar power are axiomatically perceived as a technical problem to be solved by engineers, rather than as an inevitable consequence of its dependence on great quantities of materials and fossil fuels. The fact that photovoltaic power after decades of dedicated technical research remains a privilege of the wealthy is not permitted to contaminate the vision of a miraculous technology beyond petroleum. In this way, the deliberations of engineers and economists can continue to project the illusion that they have nothing to do with environmental justice.

If solar energy was presented as an attractive option, not least for developing countries, already 50 years ago, why has its adoption remained so marginal? Although physicists, engineers, economists, and social scientists tend to address the issue from divergent vantage points, it should be important to consider their different perspectives in order to assemble a less fragmented, more encompassing understanding of the conceptual, social, and material dimensions of solar energy.

The Relevance of Thermodynamics for Economics

The prospect of a widespread shift to the direct use of solar energy is intertwined with the issue of whether the physical laws of thermodynamics significantly constrain economic processes. As previously mentioned, the economist Nicholas Georgescu-Roegen (1971) demonstrated that economic activities implicate the Second Law of Thermodynamics, as they inevitably result in a dissipation of available energy and an increase in total entropy. This observation has become fundamental to the transdisciplinary field of ecological economics. However, Georgescu-Roegen (1991: 24–29) also presented the more controversial argument that not only energy but matter, too, is inexorably and irreversibly degraded—that is, disordered and rendered less available—in economic processes. He made this claim explicitly to challenge "salvation programs" based on solar power (ibid.: 26). Georgescu-Roegen's examples include the dissipation in soil, water, and air of minerals such as copper and phosphorus, and of other refined substances such as the rubber in car tires. This argument is necessarily at odds with expectations of indefinitely continued economic growth. Critics have objected that there is no theoretical reason why the direct use of solar energy could not be harnessed to counteract the dissipation of matter by concentrating and recycling essential substances. This objection has been raised by physicists (for instance, Ayres 1999; Kåberger and Månsson 2001) as well as by Marxists committed to scenarios based on global technological progress (Schwartzman 1996, 2008). It highlights fundamental questions regarding the relations between Marxist theory,

The Political Ecology of Technological Utopianism • 121

thermodynamics, and social-science understandings of nineteenth-century technological evolutionism.

Georgescu-Roegen (1982: 10) observes that it is "beyond any question that matter dissipates primarily through friction of solids or fluids." He urges readers to think of "automobile tires, of the river banks, of the body of any living creature, briefly, of any material object with a definite form" (ibid.) and rejects what he calls the modern energetic dogma, which holds that dissipated matter can be completely recycled, if only sufficient energy can be applied. As even the economic processes that organize recycling convert available energy and matter into waste—that is, unavailable energy *and* matter—there can be no *complete* recycling of matter, regardless of the amount of energy applied. Any effort at recycling will produce additional waste. "This," writes Georgescu-Roegen (1986: 7), "is a regress without limit." The difficulties involved, he observes, are "instructively revealed by planning how to reassemble all the rubber molecules eroded from automobile tires by road friction" (Georgescu-Roegen 1982: 16). In the context of rejecting the modern energetic dogma, Georgescu-Roegen (1982: 26–33; 1986: 8–10) also dismisses various versions of an energy theory of economic value proposed, for instance, by Sergei Podolinsky, Howard T. Odum, and Robert Costanza. Finally, to illustrate that a "feasible" technology is "not necessarily viable," Georgescu-Roegen expresses strong doubts about the prospects of the direct use of solar energy (1986: 15–17; 1991: 23): in spite of "the loud din about the solution of the energy crisis by the 'cheap and renewable' solar energy," he argues that the weakness of solar radiation reaching the Earth's surface means that "we need a disproportionate amount of matter to harness solar energy in some appreciable amount." According to Georgescu-Roegen, this constraint means that solar power will never be able to satisfy the demands of high-tech society in the way that fossil fuels have.

Energy expert Smil (1992: 1–2) concedes that Georgescu-Roegen is "correct in principle," observing that biological processes dissipate both matter and energy and using expressions such as "low-entropy energies and materials." Smil also agrees with Georgescu-Roegen's criticism of an energy theory of economic value, arguing in particular with Odum (ibid.: 3–5). Although disagreeing with his use of thermodynamics, the physicist Robert Ayres (1998) sympathizes with Georgescu-Roegen's conclusion that "material dissipation does impose real constraints on the economic process." Like Smil, Ayres uses expressions such as "low" versus "high entropy materials," and agrees with the proposition that "materials can never be recycled with 100% efficiency because there are always entropic losses" (ibid.: 3). Even if Georgescu-Roegen's pessimism cannot be based on valid inferences from the laws of thermodynamics, Ayres apparently argues, the practical conditions for complete recycling would be "very hard to satisfy" (ibid.: 10). Ayres bases this conclusion on calculations demonstrating that the "wastebasket of inactive high entropy materials" would need to be very large in mass terms (ibid.: 8). The question of what is technically "feasible" versus "impossible," it seems, can be addressed at different levels of theoretical rigor. The practical impossibility of complete material recycling may not be mathematically derived from the laws of thermodynamics, but may still constitute a very real constraint on the implementation of utopian engineering.

Physicists Tomas Kåberger and Bengt Månsson (2001), like Ayres, reject aspects of Georgescu-Roegen's use of thermodynamic theory. Furthermore, they are optimistic about the direct use of solar energy to sustainably recycle material resources. Although Kåberger and Månsson concede that the concept of entropy applies both to matter and energy (ibid.: 167), and praise Georgescu-Roegen's attempt to bring "economists and economics back towards reality" (ibid.: 172), they maintain that "with appropriate technology and social organization, it is possible to slow down or even to reverse such dissipation processes, thereby building up stocks of low entropy material in society while exporting the corresponding entropy from the Earth" (ibid.: 169). These physicists, in other words, believe that technology can do what Podolinsky (2008 [1883]) thought only human labor could achieve (see chapter 5).

In a revealing discussion of the relation between entropy and economic value, Kåberger and Månsson (2001: 173–174) agree with Georgescu-Roegen that "low entropy is not sufficient for something to be valuable," citing his example that "a person may prefer an omelette to an intact egg." On the other hand, it should be observed, preparing an omelet requires the dissipation of energy, which validates Georgescu-Roegen's general point—accepted by Kåberger and Månsson—that, *taken together*, as a "necessary consequence" of the Second Law of Thermodynamics, "it is true that 'matter-energy' enters the economic process in a state of lower entropy than the state at which it leaves" (ibid.: 173). As any economic process will have produced entropy, "the more valuable products and waste materials, taken together, will have greater entropy than the total entropy of the less valuable inputs" (ibid.: 174). Although it is obvious that the economic value—that is, market price—of particular commodities cannot be expected to correlate with the amount of labor or other energy expended in their production, it would be wrong to conclude, as do Kåberger and Månsson, that "it is difficult to defend any general, meaningful statement on the relation between entropy and value" (ibid.). As these authors on the very same page have offered precisely such a statement, it is worth repeating: *The more valuable products and waste materials, taken together, will have greater entropy than the total entropy of the less valuable inputs.* Exactly this is Georgescu-Roegen's most fundamental point, but

The Political Ecology of Technological Utopianism • 123

neither he nor any other economist or physicist that I am aware of has drawn the logical conclusion that the market exchange of finished industrial products for fuels and raw materials will inexorably *reward* the dissipation of such resources with more resources to dissipate (Hornborg 1992, 1998, 2001a). In other words, the more resources we have dissipated today, the more new resources we will be able to dissipate tomorrow.

Regarding the prospects of the direct use of solar energy, Kåberger and Månsson (2001) simply declare that Georgescu-Roegen's conclusion is wrong. They envisage a future "industrial society independent of the Earth's deposits of low entropy resources, a society that, like natural ecosystems, uses the solar radiation to manage and reduce the entropy of matter," confidently claiming that "technologies are already available for running all currently fossil-fuelled processes with solar energy alone" (ibid.: 177). In effect, what they are suggesting is that there is no essential difference between biomass and technomass. In this view, the societal component of technologies is irrelevant for their capacity to mimic biological systems. For many social scientists, such statements from physicists engaged in engineering sciences sound too good to be true. To *whom* are these technologies available? If they are available to everybody, why do we continue, at increasing cost, to extract fossil fuels from tar sands and deep-sea drill holes?

Marxism, Ecological Economics, and Technological Utopianism

The technological pessimism exemplified by Georgescu-Roegen has prompted objections not only from mainstream engineers and physicists, but also from Marxists (for instance, Schwartzman 1996, 2008). This convergence is interesting, as it suggests a common confidence in technological progress, regardless of ideology. Ted Benton (1989: 55) has argued that the "bad blood" and mutual suspicion between Marxists and ecologists derives from a flaw in Marx's economic thought that makes it unable to recognize and explain ecological crises and that ultimately derives from "an insufficiently radical critique of the leading exponents of Classical Political Economy." Among Marxist notions adopted from classical political economists such as Ricardo, Benton (1989) claims, are the reluctance to admit significant natural constraints and the labor theory of value. In fact, Benton suggests that Marx can be understood as "a victim of a widespread spontaneous ideology of 19th-century industrialism" (ibid.: 61). In response to Benton, Reiner Grundmann (1991: 118–119; emphases in original) observes that Marx's contradictory approach to machine technology was "to attribute all negative aspects of machine technology to its capitalist *use*, and to attribute all positive aspects to machine technology *as such*." As argued above, it is precisely this notion of "technology as such" that social science can no longer consider tenable. It thus remains a central problem of Marxism.

With an article entitled "Solar Communism," the Marxist biologist David Schwartzman (1996: 1) hopes to dissipate a "fog of confusion" generated by Georgescu-Roegen's understanding of entropy as an "indicator of the ultimate limits of a growing economy." Although Schwartzman concedes that Georgescu-Roegen's argument regarding the relationship between entropy and the economy is applicable to an economy based on nonrenewable energy, he asserts that, given recycling and a waste-free technology, the "use of solar energy will make possible an increase in the physical throughput (material processing) in the human-made technosphere without adverse impact on the biosphere" (ibid.: 10). Referring to a source published in 1993, Schwartzman is convinced that photovoltaics "now have a bright future as a preeminent renewable energy source" (ibid.: 11) and that a solar-based economy is "a necessary condition for a global civilization realizing the Marxian concept of communism" (ibid.: 1).

In a more recent article, Schwartzman (2008: 43–44) dismisses concerns about peak oil and other natural constraints on the economy as neo-Malthusian, regressive ideologies and repeats his critique of Georgescu-Roegen's argument as based on "very shaky foundations" such as conflating isolated and closed systems. In particular, Schwartzman (ibid.: 50) criticizes Paul Burkett (2005a) for supporting Georgescu-Roegen's theory of entropy "in an apparent attempt to seek convergence of Marxist theory with ecological economics." Schwartzman's technological utopianism is evident in his dismissal of photosynthesis as a low-efficiency collection of solar radiation, and his prediction that, in the distant future, "humanity will plausibly expand outward in our solar system and even further into the galaxy" (ibid.: 52–53). He asserts that the specter of unavailable matter is irrelevant to a solar physical economy, and that tapping the "solar flux has a huge potential as the energy basis of a solar utopia" (ibid.: 53–54). In fact, it seems that the feasibility of such a solarized economy is essential to any concrete visions of communist utopia, the material prerequisites of which include a progressive dematerialization of technology through the expansion of information technology and an elimination of sprawl, leaving extensive biospheric reserves. If this technological utopia is "simply wishful thinking," Schwartzman concludes, then "any meaningful progress for humanity in this century" is unthinkable (ibid.: 56–58). This conclusion aptly expresses the existential dimension of the current impasse of technological utopianism, shared by most ideological colors of the political spectrum. It illustrates how constrained we tend to be by our focus on the tangible materiality of machines, and how blind to the possibilities of reorganizing the economic flows on which they depend.

The Political Ecology of Technological Utopianism • 125

Why Solar Panels Do Not Grow on Trees

Technological utopianism is based on a conception of technology that reflects the historical experience of core nations of the capitalist world-system. This conception envisages technological solutions as straightforward challenges of engineering, rather than as societal strategies embedded in both economics and ecology. After half a century of rhetoric on the imminent expansion of solar technology, it is high time to scrutinize this utopia in terms of its feasibility in relation to the global distribution of purchasing power and environmental degradation. Given its high costs and resource requirements, it is legitimate to ask to *whom* it will be accessible, and at the expense of *whose* resources and labor.

In an increasingly desperate pursuit of optimistic visions of a viable future for modernity, journalists have visited remote villages in Algeria, where solar panels have been installed to generate electricity for some light bulbs. The light bulbs seem to be appreciated as long as they work, even though the villagers have had to wait for years for repairs. Unfortunately, our collective dream of a technological salvation beyond peak oil tends to rest on such frail foundations. With all due respect to light bulbs, after 50 years of rhetoric about solar power it would be heartening to finally see locomotives, tractors, or bulldozers propelled by the sun.

The sun has generated billions of years of biological evolution on our planet, but why do we imagine that our species should be able to construct technologies that are more efficient at harvesting solar energy than photosynthesis? The whole idea that we could harvest direct solar energy in order to *replace human labor* should be scrutinized by the *social* rather than the technological sciences. The electricity generated by solar panels in the Sahara desert will no doubt be reserved for the people who can afford it—more likely Germans than Algerians. The same global elite, in other words, who today can afford oil. Solar technology thus seems unable to solve problems of global distribution, but there are also several difficult questions regarding its economics and ecology. Will even Germans be able to afford it in the midst of financial crisis? Where will the rare earth minerals be extracted, and with what environmental consequences? The conventional dilemmas of modern technological society appear to be able to resurface, whatever the technology.

As Georgescu-Roegen realized, there *are* also absolute physical limits to growth and resource extraction, ultimately defined by the laws of thermodynamics. The inclination of most economists and proponents of economic growth to dismiss this obvious ecological truth is remarkable. Even if no one can predict when in history or where in the world such limits will be encountered, it is obvious that they exist. From a local perspective, to be sure, it seems as if technology has made progress, but not until recently have we begun to realize the extent to which such technological progress boils down to a *redistribution* of temporal and spatial resources in global society. For instance, the historically increasing agricultural harvests that give the growth optimists such hopes for the future have primarily been based on imports of guano, phosphates, oil, and other resources from extractive sectors of the world economy. Is this resource-intensive agriculture to serve as a model for less affluent nations? The decisive question, in order for it to be rational to replace labor in one part of the world with technologies based on imports of natural resources and embodied labor from other parts of the world, is how labor and resources are *priced* in the different areas. This is why technology is ultimately a question for the social sciences, rather than engineering. We must ask, for *whom* will it be possible to invest in solar energy or household robots?

Mainstream versions of Marxism and ecological economics share a concern with the material prerequisites of economic processes, but also a tradition of analytically merging the physical and semiotic aspects of such processes in misleading ways. Rather than viewing industrial production as inherently problematic, as does Georgescu-Roegen, Marxists and ecological economists tend to envisage solutions to sustainability problems in terms of better technologies and proper pricing. In analytically distinguishing the physical and semiotic aspects of economic processes in a consistent way, Georgescu-Roegen recognizes problems of sustainability as generated by the very interaction of physical laws and the logic of money. He is thus unable to share any hopes about transcending those problems through new technologies or manipulations of market valuation, as long as humans continue to maximize monetary and consumer values in a universe obeying the laws of thermodynamics. As shall be elaborated in chapter 8, it should be obvious which of these two conditions is amenable to political change.

We may ask, finally, in what sense the existence of a given modern technology can be said to be possible in the absence of asymmetric exchange and environmental load displacement. The question might be answered by again comparing technology to organic life. It has been suggested that biological genotypes exist only as abstractions; it is only as phenotypes—actual, material processes and relationships—that organisms exist (Ingold 2000). Similarly, while a technology may seem feasible in the sense that its realization appears to be possible to achieve independently of the societal context, its viability as a mode of socioecological organization is dependent on specific socioeconomic conditions. For even a single prototype to be constructed, the inventor would need to have access to certain amounts of money, resources, and labor. More importantly, for the widespread adoption of the technology to be feasible, its metabolic properties in terms of required inputs and possible outputs would have to be aligned with the market prices of those inputs and outputs. The technology, in other

The Political Ecology of Technological Utopianism • 127

words, needs to be able to serve as a crystallization or mediator of extant market relationships. It exists by virtue of the quantitative discrepancy between flows of money and flows of matter-energy, a discrepancy that has posed an analytical conundrum to economic thought ever since the origins of money. To give a conspicuous example, a spacecraft for interplanetary travel is probably possible to construct in the very wealthiest centers of capital accumulation, but this does not mean that we can imagine space travel as a feasible future technology disembedded from the global impoverishment of people and ecosystems. Nor, in accordance with the argument in chapter 5, would it help us to expose the fetishized conception of technological progress in mainstream thought to suggest that the existence of a space shuttle in the United States of America is founded on the underpayment of biophysical "use values."

CHAPTER 8

Redesigning Money to Curb Globalization and Increase Resilience

In December 2001, the government of Argentina froze the country's bank accounts in an attempt to avoid defaulting on a debt payment owed to the International Monetary Fund (IMF). When the peso was detached from the value of the US dollar, Argentinians suddenly found their savings worth a quarter of their former value. As they were not permitted to withdraw more than 300 pesos per month from their bank accounts, millions of people resorted to barter and informally circulated credit notes in order to survive. As much as 52 percent of the population was living in poverty, and roughly 20 percent in "severe" poverty, which could include starvation (North 2007: 151). To some, it was anarchy. To others, a better economy and society was replacing neoliberalism.

In July 2015, after having defaulted on a debt payment to the IMF and found their savings frozen, the people of Greece in a national referendum voted to decline the austerity measures demanded by the rest of the European Union as a condition for further loans. Like Argentina 14 years earlier, the country was bankrupt. While Argentina could restore its financial solvency by agreeing to serve as an extractive zone at the disposal of core areas of the world-system, Greece belongs to the traditional core and has few natural resources to offer. It is ironic that the country which developed the first coins—and whose philosopher Aristotle articulated the first incisive critique of money—should also be the first country to be irrevocably disillusioned with the fantasy of money.

As in so many other financial crises through history, money reveals its fictive character when it can no longer provide for real material needs. Many categories of people can be blamed for sharing the responsibility for such economic disasters and their extensive human suffering, but ultimately financial crises raise questions about the adequacy of monetary policies and the economic worldview on which they are based. The realization that money is fundamentally a fiction has over the centuries inspired numerous critical thinkers to advocate some kind of real, material standard of value—for instance, the energy standard proposed by the Technocrat movement or the gold standard established at Bretton Woods—but such approaches do not acknowledge the dissipative character of economic processes. In this final chapter we shall discuss how an economic system might be designed which does not simply peg money to a material standard—which has always ended in failure—but which guarantees that the operation of money does not jeopardize the material security and survival of the humans that rely on it. The chapter argues that a sustainable and resilient economy will require the establishment of a complementary currency that distinguishes between values pertaining to local human survival, on the one hand, and the values in which financial institutions speculate, on the other. In order for such an alternative currency to accomplish a transformation of the economy, it concludes, we may learn from the mistakes of earlier experiments with local currencies.

The Rationale, History, and Prospects of Experiments with Alternative Currencies

Mainstream (neoclassical) and most heterodox (Marxian and ecological) economics remain confined within a worldview fundamentally shaped by general-purpose money. In not fully acknowledging the implications of Georgescu-Roegen's (1971) observations on the entropy-increasing character of economic processes, deliberations on economic policies, no matter how seemingly radical, that do not question the use of such money tend to promote increasing centralization, polarization, and environmental degradation. Although the many disadvantages of increasing scale and the obsession with economic growth were clearly articulated already in the 1970s (for instance, Schumacher 1974; Daly 1977), the conceptual lock-in of general-purpose money has continued to constrain the widespread aspiration, four decades ago, to envision an alternative emphasis on community, localized resource flows, and sustainability. Perspectives drawing on discourses on political ecology recognize that the inexorable tendencies toward globalized resource transfers, large-scale organizations, centralized power hierarchies, increasingly severe inequalities, local vulnerability, and ecological deterioration are inherent in the discourse on economics shared by mainstream and heterodox traditions (M'Gonigle 1999). But such insights from the wide spectrum of approaches here subsumed under the umbrella of political ecology only rarely identify the phenomenon of money itself as the root of all these undesirable tendencies (ibid.: 23), and even more rarely suggest an alternative.

Perspectives from heterodox schools such as Marxian and ecological economics converge in observing that monetary exchange values tend to obscure the biophysical substance of the goods and services that are exchanged. Both schools recognize that money can thus conceal asymmetric transfers of embodied labor or resources, generating polarizations and inequalities between those who accumulate and those who are impoverished. A problem identified by both schools is the inclination of mainstream economists to exclusively focus on the internal cybernetics of systems of monetary market exchange, deliberately or unintentionally ignoring causal connections between the semiotic and material aspects of economies. As Heilbroner (1999 [1953]) shows, mainstream economics has become concerned only with the logic of a monolithic market and with the systemic consequences of various kinds of policies to regulate it. From the perspectives of Marxian and ecological economics, this disregard for the substance of exchange means that important determinants of economic processes are excluded from view, surfacing only in the form of unanticipated crises. Financialization represents a decisive disjunction of the logic of money from the physical conditions of production and human life. The metaphor of a bursting bubble, frequently used in describing financial crises, illustrates that money in this form is ultimately a mere fantasy. Credit is not a matter of borrowing money in the sense of fetching it from a bank, but a promise to the bank to fulfill its fantasies of future debt service. Fantasies like these will work as long as people agree to subscribe to them, but, as financial crises have shown, when they no longer do so, money will dissolve into thin air. The volatility of cultural constructions such as the fantasy of money would not be a problem if it was not so inextricably intertwined with the material realities of human lives, from the tangible, physical metabolism of eating and working to housing and environmental impacts. For many millions of people worldwide, the recent financial crises have created severely difficult problems of a very material nature. Many heterodox economists would point out that the problems generated by the failure of mainstream economics to acknowledge material aspects of the economy are experienced by these millions of people precisely at this tangible level of reality which economics excludes from view.

Rethinking the Commensurability of Values

It is both unrealistic and futile to propose a fundamentally revised discipline of economics, which links monetary flows to flows of embodied labor, land, or energy, but it may be slightly more realistic to suggest means of insulating people's basic material needs from the vicissitudes of financial fantasies. The point of departure for the proposal to be presented here is that it is the semiotic vacuity of general-purpose money that accounts for its complete detachment from material referents and its encouragement of generalized commensurability. This universalized and increasingly globalized commensurability—the assumption that almost all values are interchangeable—is a cultural conception that ultimately jeopardizes not only human civilization but even the biological conditions for human life. To curb the destructive societal and ecological processes currently generated by the phenomenon of money, it will be necessary to redefine our cultural conception of commensurability. Such a shift means distinguishing values pertaining to basic human survival from the values in which financial institutions speculate. This would not need to be a matter of legislation as it would suffice to provide people with other options for survival than to sell their labor and buy their food in the same market that is used by corporations as an arena for capital accumulation. If people would indeed tend to prefer the alternative option, a fundamental transformation of the global economy could conceivably occur without either legislation or coercion. The idea is for national authorities to issue a complementary currency,[1] which can only be used to purchase locally produced goods and services, and to distribute it as a basic income to all households in proportion to their size. To define what is to be categorized as locally produced, a reasonable procedure might be to restrict the use of this complementary currency (let us provisionally call it Points) to purchases of goods and services originating within a given radius (say, 30 km) from the place of purchase. A practical way of distributing Points to households would be to provide them with plastic cards which are automatically charged with new, electronic Points each month, in the same way that credit cards give access to salaries. It will immediately be recognized that this proposal deviates in important respects from the many experiments that have been conducted with so-called local or community currencies in various parts of the world. Before discussing its advantages, we shall briefly review some recurrent features of these experiments.

The widespread recognition that the growing dependence of local communities on the global market economy has had a number of unfavorable repercussions—such as greater vulnerability and disempowerment, loss of social cohesion, and the exploitation of local labor and resources by distant centers—does not need to be reiterated. The idea of countering such processes by resorting to a local community currency has emerged in various places and at various times. It was widely discussed in nineteenth-century Europe and the United States, and several social movements attempted to implement it (North 2007: 41–61). The most well-known modern movement toward this goal is the ambition, beginning in Canada and the United Kingdom in the 1980s, to establish the so-called Local Exchange (originally Employment) Trading Systems (LETS) (Dobson 1993; Douthwaite 1999; North 2007), but similar initiatives have appeared in Argentina, Australia, Austria, Belgium, Czech Republic, Germany, Greece, Hungary, Japan, New Zealand, Poland, Slovakia, Sweden, United States, and several other countries. In some cases—most conspicuously Argentina at the turn of the millennium and more recently Greece—the idea of complementary currencies emerged as a survival strategy and as an explicit response to severe financial crisis.

These movements have become a field of academic study with its own journal, the *International Journal of Community Currency Research*. A special issue (Blanc 2012) provides a recent overview of the history and prospects of such experiments with alternative currencies. Recurrent shortcomings include widespread dismissal, absence of a national governance system, inefficient promotion of local consumption, personal exhaustion of leaders, insignificant impact, accounting difficulties, risks of free riding, and unclear incentives on the part of shopkeepers. The editor concludes that "thirty years after their first emergence, [community currencies] still have to prove they can change the present state of things, while research agendas are increasingly considering them" (ibid.). According to another assessment, LETS are now on "a worldwide retreat" (Dittmer 2013: 6). However, the shortcomings revealed by systematic research on these movements provide a foundation for designing a complementary currency system that is fair, widely utilized, government regulated, easily administrated, and efficient. A key challenge is to design this system in such a way as to provide all significant social actors—households and businesses as well as authorities—with strong incentives to participate.

The predominant justification for most complementary currency systems that have appeared so far is that they represent "forms of micropolitical resistance" from below (North 2007: 77). This means that they are generally grassroot initiatives largely contingent on the enthusiasm and ideological commitment of a restricted number of activists, with little or no support from authorities (Dittmer 2013). It also means that they are unlikely to reflect systematic analysis of the conditions under which they might succeed, including considerations of fairness, attractiveness, large-scale administration, efficiency, impact, and transparency. The system that is advocated here differs from most of these initiatives in the following respects: (1) It would be organized by the federal or municipal authorities. (2) The currency (Points) would be distributed by the authorities as basic income to all citizens. (3) The Points would only be useful for purchases of local goods and services, that is, goods and services originating from within a specified radius from the place of purchase.[2] (4) All transactions with Points would be officially exempt from taxation. (5) To the extent that some individuals wish to save Points for later use, while others may temporarily want to borrow extra Points, special institutions would administrate such electronic transactions, but without offering or charging any interest. (6) Businesses would have the option of converting a portion of the Points they earn into regular currency, through the authorities, at adjustable rates calculated to compensate for the authorities' loss of tax revenue. (7) Parts of the authorities' expenditures for pensions and social security would be paid in the form of Points. Under these conditions, all significant social categories would benefit from the Point system.[3]

By systematically considering this arrangement from the perspectives of the social actors concerned, it is possible to avoid most, if not all, of the disadvantages and shortcomings of LETS and related community currency systems. Households would be able to liberate some of their regular income by utilizing Points, whenever possible; they would also be less dependent on salaried work and less vulnerable to unemployment; finally, they would experience more local interdependence, cooperation, and sense of community. Businesses would find opportunities for tax-free income, some of which could be used to purchase local resources, some to flexibly employ local labor, and some to convert into regular currency; there would also appear new opportunities for diversified local enterprise to satisfy the increasing demand for a wide range of local goods and services. Authorities would reduce their costs for pensions, social security, medical care, transport infrastructure, and environmental protection, thereby avoiding risks of fiscal deficits. Some of the many societal benefits of this system are: lower demand for long-distance transports (reduced greenhouse gas emissions, energy use, transport costs, and traffic accidents); more local recycling of nutrients and packaging materials (reduced eutrophication, solid waste, and resource depletion); less mechanized agriculture (reduced resource use and environmental degradation, more physical exercise for significant parts of the population); lower demand for export production of food (globally reduced vulnerability of rural populations, increased self-sufficiency, and food security); more localized food production (less waste through overproduction, storage, and transport; fresher and healthier food with less preservatives; better transparency in relations between producers and consumers); more diverse landscapes (higher biological diversity and ecological resilience); more diversified local business profile (demand for a wide range of local goods and services); greater financial resilience of federal governments (lower costs for pensions, social security, and other major expenditures); and more social cohesion (less social marginalization, more sense of community, and better psychosocial health). All these benefits could be achieved by establishing a complementary currency thus designed, enhancing financial, social, and ecological resilience while not constraining the global market from encouraging vital industries—such as advanced medical equipment, pharmaceuticals, and information technology—that would continue to be in demand and require global integration. The advent of electronic money in 1971 certainly unleashed an unprecedented fetishization of the global economy,[4] but it also opened up completely new possibilities to design currencies that promote equality, democracy, and sustainability (Hart 2000). Two thousand years ago, St. Paul was no doubt right in that money is the root of all evil, but at this point in history Bernhard Lietaer (2001: 7) is also right in that it is "the root of all possibilities."

Electronic money has a potential for making the economy more sustainable and equitable for the same reason that it has promoted financialization and financial crisis, that is, its *lack of material form*. Following the delimitation of its ideal use articulated by Aristotle, money should merely be a medium of exchange between socially connected producers and consumers. It should be a means, not an end in itself. But money inevitably becomes an end in itself when it is attributed with *intrinsic* value, as when precious metals or bills are hoarded or stolen, or when interest accrues on bank accounts. This is money fetishism. However, money that is both electronic and interest-free has no intrinsic value. In this form, it can finally serve its makers, rather than make them its servants.

The fundamental goal of a complementary currency system such as sketched here is to *relocalize* much of the material metabolism of human societies, essentially because such a strategy is both more equitable and more sustainable than current trends. This is "the precise opposite of the modern trend of globalization" (Lipson 2011: 573; cf. Brennan 2003). In Marxian terms, it would mean an expansion of simple commodity circulation (C-M-C^1) at the expense of capitalist circulation (M-C-M^1) and financialization (M-M^1). It would not require violent revolution, but merely the existence of an option that would be attractive and sensible to everybody. In fact, it would not even mean abandoning the insight of mainstream economics, from Adam Smith onward, that market exchange is an efficient way of allocating resources, because it does not challenge the market principle as such, only the *scale* of market organization. The chances of achieving the hypothetical perfect information imagined by economists inevitably diminish with increasing market scale. Nor could this proposal for a relocalization of the market be dismissed as regression, as it would be based on recently emerging, transdisciplinary understandings of economic processes and on new digital technologies. History is not reversible, but we can take stock of millennia of historical experience in order to envisage our future.

The Shortcomings and Insights of Resilience Theory

It can be argued that discourses on the sustainability of human–environmental relations that ignore their political dimension are not only incomplete, but in themselves—as ideologies—manifestations of power. The currently burgeoning discussions on social–ecological resilience (Berkes and Folke 1998; Levin et al. 1998; Gunderson and Holling 2002; Berkes et al. 2003; Folke 2006) tend to mask the power relations, contradictions of interest, and inequalities that to a large extent determine how humans utilize the surface of the Earth. On the other hand, resilience theory has an underexplored potential to radically confront such power structures by identifying some of the basic assumptions of economics as the very source of vulnerability, mismanagement, and crises. As has been argued in this and previous chapters (see particularly chapter 3), the most basic assumption of economics is its faith in general-purpose money and global markets as signaling systems that promote the most efficient allocation of resources. Contrary to this assumption, the logic of general-purpose money in several respects promotes *in*efficiency, if other parameters such as energy are taken into account. Of more immediate relevance here, however, is the inclination of general-purpose money and global markets to reduce local socioecological resilience. This conclusion can be derived from the systems-theoretical tenets of resilience theory itself. These tenets can be used to argue for special-purpose currencies and local markets that complement the global economy, as sketched above, rather than an undifferentiated globalization of resource flows. The ultimate implications of resilience theory, in other words, are vastly more radical and subversive than its current proponents imagine.

The emergence of resilience discourse in recent years has been critically discussed from several angles, tracing its intellectual ancestry in systems ecology, its ideological affinities with neoliberal economics, and its incapacity to account for actual patterns of land use in various parts of the world (Hanley 1998; Lélé 1998; Brand and Jax 2007; Gotts 2007; Nadasdy 2007; Hornborg 2009; Kirchhoff et al. 2010; Park 2011; Walker and Cooper 2011; Reid 2012; Sheridan 2012; Widgren 2012). It has also been scrutinized microsociologically as a social movement explicitly determined to avoid criticism (Parker and Hackett 2012), which raises questions about its solidity as a scientific endeavor.

The use of the concept of resilience in public and academic discourse on human–environmental relations reflects an ideological assimilation of environmental concerns by an establishment keen to avoid alarmist messages challenging business as usual. Although represented as a synthesis of perspectives from both the natural and the social sciences, resilience discourse generally appears to be ignorant of most of the tenets of modern social science, except for occasional contributions from economists eager to develop new mathematical models for natural resource management. Leading advocates of the resilience of traditional resource management, for instance, reveal a very superficial grasp of anthropology (Berkes and Folke 1998; Berkes 1999;

Berkes et al. 2003). Their proposal that the modern concept of social–ecological systems has affinities with traditional ecocosmologies is highly misleading, as is obvious from the fact that the latter tend to extend the social domain into the natural, rather than vice versa (Descola 1994). Passing references to what Lévi-Strauss (1966) has called savage thought (Berkes and Folke 1998: 12–13) similarly miss the point of his analyses entirely. Even more problematic than its distortions of anthropology, however, is what Lélé (1998: 253) identifies as its pervasive inattention to "major asymmetries in the interests and powers of the different actors." The "panarchical perspective," writes Gotts (2007: 6), "has had little to say about social elites and the often violent and oppressive ways in which they maintain themselves." This conspicuously ideological dimension of resilience discourse prompted Nadasdy (2007: 217–218) to conclude that it "has the implicit goal of maintaining the social–ecological relations of capitalist resource extraction and agro-industry."

The key metaphor of the resilience movement is the model of the adaptive cycle applied decades ago by the ecologist Crawford Holling to forest ecosystems in eastern Canada. The famous horizontal figure eight recurs in countless publications on resilience theory, including the cover of the canonical volume *Panarchy* (Gunderson and Holling 2002), and seems to be the principal common denominator of resilience research. As such, its many uses deserve special scrutiny. Social scientists are commonly disturbed by claims that the model of the adaptive cycle is applied not only to ecosystems or social–ecological systems, but even to social systems as such. Resilience theorists have thus proposed analogies, for instance, between old-growth forests and large corporations, and between forest fires and financial panics (Peterson 2000). Apparently, the growth phase of the adaptive cycle is as applicable to forest biomass as to the logging industry, or to fish stocks as to fisheries. This obviously represents a logical contradiction when applied to social–ecological systems where the growth of societal capital is inversely related to the growth of natural capital, for instance, where the growth of the logging industry is associated with the depletion of forest biomass, or the growth of fishing fleets with the depletion of fish stocks. This contradiction can be resolved only by concluding that the social and ecological components of such systems follow separate and antagonistic cycles, where the growth phase of economic capital coincides with the release phase of natural capital. But if social and ecological systems follow distinct and contradictory cycles, it no longer seems meaningful to conceive of them as components of a single system tracing a common adaptive cycle.

It has been observed that the concept of resilience can be more or less precisely defined in engineering and ecology, but can serve only as a vague and contested metaphor in the social and behavioral sciences (Brand and Jax 2007), where it will inevitably raise normative questions about the relative desirability of different states and conditions. In order to discuss the resilience of a particular social–ecological system, it would be necessary to define its geographical boundaries, its exchanges with the world outside those boundaries, its physical constitution (population, resources, metabolism, etc.), its social and cultural organization, the relevant physical and social parameters and their acceptable ranges of variability, the vulnerability of these parameters to disturbance, and so on. In spite of the voluminous rhetoric on the resilience of social–ecological systems, there has been no case study in which all these aspects have been competently addressed. It is doubtful if it is at all feasible to do so.

An example of the kind of quandaries that theorists of societal resilience will run into is the problem of how to define its opposite, that is, societal collapse. The concept has been defined by anthropologists and archaeologists as a sudden loss of political or economic integration and complexity, resulting in a fragmentation into less inclusive and more autonomous social units (Tainter 1988; Yoffee and Cowgill 1988). World history can be viewed as sequences of such collapses, followed by periods of greater local autonomy and then renewed integration, but is it correct and meaningful to approach world history as an adaptive cycle? What does it mean to say that civilizations are complex adaptive systems? Can social systems learn?[5] At which level of social inclusiveness are the resilience theorists concerned with resilience? Was the collapse of the Roman Empire a failure of resilience, or was the survival of some postimperial communities an index of adaptive success? Is the contemporary world market a similar social project destined for collapse, or is the preoccupation of neoliberal economists with resilience an indication of their commitment to preserving it? If the collapses of past civilizations are to be approached in terms of social–ecological cycles, we again need to ask why the growth of societal capital (cities, temples, roads, etc.) tends to be inversely related to the growth of natural capital (forests, topsoil, biodiversity, etc.), illustrating that social and ecological systems follow separate and contradictory cycles.

A further dilemma for systems ecologists addressing social systems is how to handle the concept of identity (Brand and Jax 2007: 4). The identity of an ecological system is a matter of objective properties remaining within a certain range, whereas social or cultural identity refers to the subjective experience of groups of people. Thus, for instance, archaeologists have seriously questioned if there really was a Maya collapse in the tenth century, as there is still a large population of people who speak a Maya language and identify with Maya culture (McAnany and Yoffee 2010). Although the appropriate response to these archaeologists would be that it is essential to distinguish between an objective historical loss of sociopolitical complexity and the persistence of a sense of ethnolinguistic identity, the ecologists' explicit suggestion to focus on the concept of identity as what resilience is all about raises the question of how the concept is to be used in transdisciplinary approaches to social–ecological systems.[6]

The theorizing on social–ecological systems by systems ecologists shows remarkably little respect for social-science research, and it is difficult to imagine examples of inverse colonization of natural by social science, as if, for instance, political scientists should begin conceptualizing ecosystems in terms of power structures. The ecologists' theory of society tends to strike social scientists as naïve and generally at odds with elementary social theory established decades ago. It appears to conceive of disasters such as societal collapse, epidemics, starvation, and war as adaptive phases of more or less natural cycles. Most remarkable is its neglect of decades of voluminous discourse on political ecology (Blaikie and Brookfield 1987; Peet and Watts 1996; Bryant and Bailey 1997; Keil et al. 1998; Paulson and Gezon 2005; Biersack and Greenberg 2006; Peet et al. 2011). This neglect is evident even in contributions that explicitly propose to integrate political ecology and resilience theory (Peterson 2000). Several authors have also explicitly contrasted the approaches of resilience theory and political ecology, arguing that the latter is better equipped to account for actual patterns of land use and resource management (cf. Sheridan 2012; Widgren 2012).

The currently expanding dominance of resilience theory within the field of sustainability research can thus not be accounted for in terms of analytical progress, but appears to reflect its capacity to ideologically defuse the challenges posed by political ecology and other conflict-conscious approaches to human–environmental relations. It remains to be shown how power might be addressed within the framework of resilience theory, so as to exploit the potential of Holling's systems ecology as a *subversive* analytical framework. These suggestions are offered in response to recurrent assertions by proponents of resilience theory that they are indeed very concerned with issues of power, but have simply not yet turned their attention to them. If the inattention to power is indeed a glaring lacuna in theories of socioecological resilience, we must encourage resilience theorists to seriously engage the topic.

Toward an Understanding of Power in Social–Ecological Systems

The resilience theorists are undoubtedly right in observing that social and ecological systems are geared to each other, or "coupled" (Berkes and Folke 1998; Gunderson and Holling 2002; Berkes et al. 2003). With this established, we must conclude that a theoretical framework capable of accounting for the dynamics of such coupled systems will have to accommodate essential elements of both social and natural sciences. On the one hand, an indispensable element of modern social theory is the pivotal role of *culture*, understood as socially negotiated systems of meanings. Cultural systems of meanings can also be referred to as symbolic systems. Without reckoning with the specific cultural ways in which the world is perceived by the particular category of humans under consideration, there can be no social theory. A central aspect of culture, of course, is language. The way in which humans collectively classify phenomena in society and nature influences their behavior in fundamental ways. On the other hand, an account that accommodates the objective realities investigated by natural sciences must also be capable of reckoning with material factors such as flows of energy and materials. In other words, a theory of socioecological systems must be able to deal with both cultural and material phenomena, that is, with flows of signs as well as flows of matter and energy.

Power is a hybrid phenomenon involving both cultural and material aspects. A general definition of power could be built on the observation that it universally implies unequal access to material resources of some kind, including energy. But in order to be complete, such a general definition would also have to account for how such inequalities of access are socially maintained. A reasonable proposition is that the most pervasive, yet least salient, way in which social inequalities are maintained is through cultural mystification, that is, by rendering them either invisible or self-evident and natural. This is simultaneously a quite concise way of defining ideology. Of course there are other means of reproducing inequalities as well, notably coercion, but it can be argued that they are generally secondary to the power of ideology. The subtle power of culture or ideology tends to be encoded in our basic and seemingly self-evident categories of thought, that is, our language.

Two examples will suffice to illustrate this point. The first is derived from the Inca Empire of early sixteenth-century Peru, where the Inca emperor persuaded millions of his subjects to invest their labor in agriculture, public architecture, warfare, and manufacturing by claiming to be the son of *Inti*, the Sun God. He and his relatives were able to claim a significant proportion of harvests and other produce by representing the unequal exchange of labor and resources as a reciprocal exchange between the emperor and his subjects. The material flows of resources hinged on the semiotic flows of words through which the former were conceived and organized, for instance, the *mita* labor tax and the *ayni* rituals, in which peasants worked in the emperor's land in exchange for maize beer. Needless to say, the volumes of maize that had been brewed into the beer served at the *ayni* represented only a tiny fraction of the maize that was harvested. The power of the Inca emperor, in other words, consisted not only of flows of matter and energy converging on his many warehouses, but also on the cultural concepts (such as *mita* and *ayni*) through which the metabolism of the empire was reproduced.

The second example is much closer to home and will thus be more difficult to assimilate, but the basic argument is the same. For more than two centuries now, Europeans and their overseas dependents have learned to find it quite natural to sell their labor time and natural resources on the market for money. They have contributed to European factories and industrial machinery, built cities, fought wars, and produced commodities. The urban-industrial infrastructures that illuminate Europe on satellite images of nighttime lights indicate vast investments of labor time, energy, and materials. Once again, we can observe that these asymmetric flows of matter and energy would not have occurred without the semiotic flows of words by which they were orchestrated, for instance, concepts such as *wage* and *market price*.

A conclusion we can draw from these two examples is that cultural systems of meanings, encoded in language, are essential components of any social arrangements for the distribution of material resources. Inequalities in social power tend to boil down to inequalities of access to such resources, including armies with which to assert them, all legitimized by hegemonic discourses, whether concerned with the divine ancestry of the Inca emperor or with the invisible hand of the market. The phenomenon of social power includes not only unequal access to resources, but also unequal influence over the construction of mainstream discourse. To understand social–ecological systems, it is absolutely necessary to address the political dimensions of such cultural discourses. In the modern world, this means addressing the political dimensions of mainstream economics.

The Subversive Implications of Resilience Theory

Let us now turn to a central observation in Holling's framework for understanding the operation and viability of living systems: the ideal congruity of temporal and spatial scales (Gunderson and Holling 2002). It appears that resilience in natural systems is importantly geared to the tendency toward a general correspondence between level of integration and longevity, so that, for instance, a forest is more permanent than a tree, a tree more permanent than a leaf, and so on. This nested, hierarchical character of living systems safeguards the more inclusive systems from being jeopardized by the failures of subsystems, and in some contexts vice versa. The relative autonomy of subsystems vis-à-vis lower or higher levels can thus be regarded as a key principle for resilience and sustainability. Societies need to be able to survive the demise of individual organisms, and organisms need to be able to survive the failures of individual cells, but individual trees inversely need to be able to survive forest fires, and local communities need to survive the collapse of empires or global markets. Such cybernetic insights long ago prompted the anthropologist Roy Rappaport (1979: 145–172) to define maladaptation in terms of communicative failures such as oversegregation, hypercoherence, and usurpation. The general understanding of socioecological crises as crises of communication can be traced to Rappaport's mentor Gregory Bateson (1972), who pioneered the field of cybernetics, or systems theory. If stripped of some of the metaphysical confidence in diffusely organized wholes, which suggests significant affinities between systems ecology and neoliberal economics, the approach is of undeniable relevance for the continuing deliberations on sustainability.

It is remarkable that resilience theory has not proceeded from such central insights of systems ecology to critically scrutinize the operation of communicative mechanisms in the modern world, the most fundamental and pervasive of all of which is money. It would be completely in line with Holling's emphasis on the principle of congruity of temporal and spatial scales to observe that what economic anthropologists refer to as general-purpose money systematically defies that principle, by making all kinds of values commensurable, regardless of which level of scale they pertain to. Goods and services pertaining to the reproduction of individual human organisms, such as food and beverages, for instance, are considered interchangeable on the world market with goods and services pertaining to the reproduction of entire ecosystems, or even the biosphere, such as technologies for deforesting Amazonia. Due to the logic of general-purpose money, people thus routinely trade rainforests for Coca-Cola.

In resilience discourse, the neglect of the destructive implications of general-purpose money is closely related to the neglect of those of global systems of exchange. In the worldview of leading resilience theorists, "social systems can be as small as a family or as large as a nation" (Gunderson and Holling 2002: 107). Presumably, the nation represents the most inclusive social system conceivable for these theorists. Gotts's (2007: 7) observation that "world-systems analysis could strengthen work within the resilience conceptual framework" is thus clearly an understatement. General-purpose money has historically extended the reach of long-distance trade and is the cornerstone of today's increasingly globalized markets. Although celebrated in neoliberal ideology, economic globalization undeniably increases the dependency and vulnerability of local communities. "In recent centuries," writes Gotts (ibid.), "largely European-derived changes in transport, communication, and military technologies have drastically reduced the autonomy of regional-scale systems." Yet, although the globalization of the market has thus increased vulnerability and reduced resilience, neoliberal advocates of this very globalization now present themselves as champions of resilience (Walker and Cooper 2011; Reid 2012).

Although incapable of analyzing the cultural and political dimensions of sustainability, Holling's understanding of hierarchies of spatiotemporal scales in living systems provides an incisive analytical tool for identifying modern money—through its capacity to confuse scales—as a source of environmental degradation. Although the tendency of money to promote the interchangeability of values at very different spatial and temporal scales obviously dissolves socioecological resilience, resilience theorists regrettably continue to approach capital as if it was as natural as biomass and markets as if they were a kind of ecosystem.

Even if the most well-meaning advocates of sustainability are thus unable to discern the cultural peculiarities which appear to propel global society toward disaster, this is not because there are no alternatives. As proposed above, it is quite conceivable to organize an economy with separate and incommensurable currencies for different kinds of values. Not only are there plenty of ethnographic examples of such multicentric economies documented by economic anthropologists such as Paul Bohannan (1955), but recent financial breakdowns in countries such as Argentina and Greece have invariably prompted initiatives in the same direction (North 2007). The creation of local currencies for community cooperation and survival is a recurrent theme in the turbulent history of international finance, and is a central component in the Transition Towns movement. The myriad ephemeral experiments with the so-called LETS illustrate the attractiveness of the idea of local currencies, but in order for this idea to generate a decisive and general break with the destructive logic of modern money, it needs to be grounded in more profound analysis and to be backed by national authorities. It is possible that the current financial turmoil in Europe and North America might provide opportunities for serious discussions about how spheres of economic commensurability could be redefined in the interests of financial, social, and ecological resilience.

Redefining Commensurability in the Light of Resilience Theory

A common denominator of the various experiments with LETS is the ambition to create an alternative, informal economy alongside the formal economy based on state-issued currencies. The main point has generally been to increase local interaction, local economic diversity, and local control over resources. Crucially, however, the alternative local currency in these experiments does not distinguish between local and nonlocal products, that is, values representing different scales of socioecological inclusiveness. In accordance with the analysis above, this ought to be the central function of a currency system that would enhance local resilience. In order to achieve the desired effects, the new local currency would need to (1) offer consumers a superior alternative to purchasing commodities with regular money and to (2) specify the range of local goods and services that it can be exchanged for. In other words, a political decision to implement such an alternative economic system would need to include strategies for (1) persuading consumers to actually use the local currency, rather than regular money, and for (2) ensuring that its use actually promotes consumption of local goods and services.

It is not inconceivable that, a few decades from now, financial or ecological crises might induce currently affluent nations to seriously consider such strategies. The potential benefits of localizing economies are not restricted to biophysical consequences such as reducing transports, energy use, carbon dioxide emissions, and waste, or enhancing biodiversity through more complex patterns of land use, but would include reducing federal expenses, for instance, for transport infrastructure, environmental protection, health services, and social security, that is, precisely the kinds of public expenditures that are already proving a heavy burden for many welfare states. Over the long term, such localization would reduce marginalization and vulnerability to various kinds of crises, enhancing cooperation, diversity, and general resilience at the local level. This is not to advocate a sudden abandonment of highly specialized, vulnerable, or disaster-prone communities, denying them the supralocal relief that they have grown used to, but to encourage a long-term increase in local self-sufficiency, autonomy, and social–ecological calibration. Carefully implemented and monitored by the federal authorities, the reform would proceed at a pace that would not risk jeopardizing human or ecosystem health.

If these benefits were acknowledged, and the authorities persuaded that such a bicentric economy would indeed relax the pressure on fiscal resources, they might find that the most efficient way of reorganizing the economy in this direction would be to electronically distribute a certain sum of the new, special-purpose currency to all citizens, each month, adjusted to their age. Assuming that households will wish to economize with their various resources, we can expect them to employ the new currency in purchasing potentially local produce such as food, clothing, and building materials, and local services such as childcare, carpentry, and repairs, because in doing so they would be saving some of their regular income for necessarily nonlocal expenditures such as information technology and pharmaceutical products. Over the long term, this shift would significantly reduce the demand for long-distance imports of commodities that can be locally produced, which should be very much in line with agendas for resilience, sustainability, and any nonmonetary measure of efficiency. As we have seen, the list of potential advantages of such a shift is long, particularly if we include not only ecological and financial but also social and existential benefits. As the new currency would be distributed to households without any required reciprocation, it could be regarded as a kind of basic income that guarantees a minimum level of subsistence even where no formal employment is available, or even desired.

The aim of such an intentional relocalization of social metabolism—that is, flows of matter and energy—would be to generate a multitude of spatially restricted but overlapping spheres of exchange, in which the average transport distances of goods and services are significantly reduced. The idea would not be to create separate local currencies for separate, bounded communities, but to allow the rationality of the single new currency to work out its own spatial-metabolic logic in terms of overlapping geographical fields of distribution. The new currency signifying local consumption would be possible to use anywhere in the country, but only to buy goods and services that are locally produced. The assumption is that it would generate incentives to both consume and provide goods and services with as short transport distances as possible. Instead of visualizing communities as bounded cells, we might anticipate their metabolic flows more as intersecting local networks. If transport distance could once again be expected to increase a commodity's price, as in the prerailway world of von Thünen (1966 [1826]), localization and diversification of production would be encouraged through market competition. Until such price competition itself suffices to promote the local economy, the determination of which goods and services qualify as local could conceivably be organized in different ways, but it might initially involve some kind of certification system specifying, for instance, a maximum number of transport kilometers for different products sold at a given market, or a range of neighboring municipalities from which they may derive. The certification system would also need to consider production methods, so as to prevent entrepreneurs from using the formal currency to purchase remotely derived inputs, such as diesel, to produce goods or services for local markets.

The localizing consequences of dividing the market into two separate spheres of exchange will be recognized as in some respects running counter to developments that we for centuries have learned to celebrate as progress and modernization. In these respects, the suggested reform would appear to be a step backward, but this is an illusion building on our cultural definition of progress. A material localization of the economy would not contradict a continued communicative globalization. The modern intensification of energy use, long-distance transports, and mechanization represents the historical experience of a privileged segment of global society over the past two centuries of fossil-fueled capitalism. From a global perspective, human progress should be defined not in terms of capital accumulation but of enhanced conditions for harmony, sustainability, health, communication, and security. In short: resilience. But rather than inspire resistance to the neoliberal world order against which it was launched, the concept of resilience has been incorporated as a central component of the neoliberal model itself (Walker and Cooper 2011). The subversive implications of Holling's recipe for resilience, which should provoke a serious confrontation with some foundational assumptions of contemporary economic theory, have yet to be explored by his followers.

A very important assumption, not only of economics but of modern thought in general, is that technological capacity should be viewed primarily as a progression in time, rather than something that is unequally distributed in social space. The uncritical subscription to this assumption is reflected in the way technology is discussed by resilience theorists, that is, as a means of extending "the ambit for human choices from local to regional to planetary scale" (Gunderson and Holling 2002: 101), rather than as a means for *some* humans to extend their ambit at the expense of others.[7] Access to modern technology is tantamount to relative purchasing power, and the rationality of any technological system is inextricably geared to relative prices of labor and resources on the world market. Thus our cultural faith in money and the market is as essential to the material accumulation of technology as was the cultural faith in the Sun God to the material accumulation of terraces in the Inca Empire. Modern technologies cannot be made universally accessible to all humans, but represent social strategies for redistributing time and space in global society. In a long-term intergenerational perspective, moreover, the contemporary extension—by means of fossil-fuel technologies—of the ambit of affluent humans occurs at the expense of future generations whose ambits are curtailed by exhausted oil reserves and climate change. The extent to which the accumulation of modern technological infrastructures is contingent on structures of market exchange would become very apparent if the reform sketched above were to be implemented. If the demand for long-distance imports was to subside, there would simply be no incentive to maintain massive infrastructures for transporting foodstuffs, energy, and materials across the globe.

A predictable objection to the vision of a bicentric economy would be that the proposal is in line with the neoliberal strategy to urge states to relinquish their responsibilities for the well-being of their citizens. It is easy to sympathize with the critiques voiced by Duffield (2008) and Reid (2012: 67), who argue that mainstream policies advocating community-based self-reliance may simply have the purpose of shifting the burden of security from states to people, and I fully share the underlying conviction that it should be the responsibility of democratically elected authorities to safeguard as far as possible the health and security of the people under their jurisdiction, which certainly means intervening in the polarizing logic of the capitalist market. Paradoxically, with the experience of Greece fresh in memory, this must now mean dismantling parts of the citizens' dependence on the financial solidity of their governments by delegating the practicalities of basic provisioning to local markets. The problems of sustainability and security currently facing humanity require analyses and policies that transcend the conventional confrontation between right-wing advocates of the market and left-wing advocates of state intervention. Unfortunately, the left-wing vision of a world of universally affluent, technologically advanced, and egalitarian nations is no longer credible. The Scandinavian countries of the 1960s, which for many served as a model for development, now represent a privileged corner of the world, blessed by the success of their export industries on the very capitalist world market that they pretended to transcend. The levels of consumption—not least of fossil fuels—enjoyed by average Scandinavians are neither physically possible to universalize among seven billion humans nor are they defensible from the perspectives of global sustainability and climate change. Ambitious federal welfare programs are feasible only as long as domestic export industries do not relocate to countries with lower salaries and lower taxes, or as long as welfare and consumption can be financed through credit. The recent financial crises in the United States and Europe indicate that the capacity of developed nations to maintain a high and preferably rising standard of living for a majority of their population is seriously constrained not only by the logic of the capitalist world economy, but ultimately also by the finiteness of the biosphere of which it is a part.[8] What the experiences in Greece should be telling us is that the cornucopian worldview of modernity is fundamentally flawed. Like all imperial strategies, financial imperialism has its winners and losers.

It is no doubt true that concerns over resilience and sustainability have been co-opted by the very neoliberal model which prompted them to emerge in the first place (Reid 2012: 74), but to dismiss ecological and/or financial concerns as neoliberal mystifications is to deny real structural problems that adhere not only to modern capitalism, but to the cultural phenomenon of general-purpose money itself. To believe that some version of socialism would enable seven billion humans to adopt the technological comforts and levels of consumption currently enjoyed by average Americans or Scandinavians is almost as naïve as Schwartzman's (2008: 53) vision that, in the distant future, "humanity will plausibly expand outward in our solar system and even further into the galaxy." Thinking realistically about the prospects for a sustainable and more egalitarian world society means navigating between the Scylla of ruthless neoliberalism and the Charybdis of high-tech utopias of solidarity. Both programs presuppose that the money and resources will be there to distribute, and that the critical issue is how to distribute it. Neither program seriously considers the possibility that faltering core areas of the world-system such as the United States and Europe will be unable to maintain their positions as regions of privileged purchasing power and undiminished mass consumption. In not having grasped the conditions for their own feasibility, both these political visions would in due time lead to socioecological disasters and unprecedented human suffering. It is against this background that we have reasons to seriously consider the vision of a bicentric economy.

This admittedly wide-ranging discussion has recommended the proponents of resilience theory to engage more respectfully with social science, particularly its understandings of culture and power. They were also advised to establish a critical distance to the metaphysical assumptions of complex adaptive systems theory, particularly when applied to social systems. Furthermore, if resilience theorists are sincerely concerned about sustainability, they have every reason to critically scrutinize the operation of general-purpose money, the global market, and neoliberal ideology. Upon doing so, they would no doubt find the idea of a bicentric economy, as sketched above, entirely consistent with the fundamental insights of resilience theory.

To conclude with one more glimpse from the ancient Andes, we can reflect on the historical fate of the local village communities (*ayllu*) that in the early sixteenth century were the building blocks of the Inca Empire. Although the emperor and his court were obviously adept at extracting surplus from these communities, they must have been granted a significant measure of autonomy, or so many of them would not have survived the traumatic collapse of the empire, followed by centuries of colonialism and impoverishment. Many rural, Quechua-speaking communities in Peru still today practice sustainable subsistence agriculture on terraces constructed several centuries before the rise of the Inca Empire.[9] The local, socioecological building blocks of precolonial Andean civilizations were apparently sufficiently autonomous to recover from the recurrent shocks of suprasystem breakdown. It is very doubtful if modern communities in Europe or North America are similarly resistant to wider systemic crises. Specialization and dependency increase vulnerability, which is tantamount to reducing resilience. We thus expect the resilience theorists, following the implications of Holling's observations, to focus their attention on how the very foundations of current economic policies need to be radically reconsidered.

Instead of fetishizing a monolithic "system" called capitalism, requiring thousands of volumes of theoretical analysis and debate to comprehend, let us talk about the fairly simple but inexorable logic of money. Marx identified this logic as M-C-M^1. If we combine the insights of Marx and Georgescu-Roegen, we realize that the production of commodity C has implied an increase in entropy, corresponding to the increase in monetary value signified by M^1. The logic of money is such that any component of the production process that can be given a monetary price, including labor, will be commoditized for appropriation on the world market. This is what has happened historically, and we know the consequences for human society and the planet. The only feasible and sustainable way to domesticate this development is by redesigning money in the direction sketched in this chapter. Whether this would justify talking about a shift toward a postcapitalist "mode of production" is less relevant than its many social and ecological benefits.

Conclusions: Money, Technology, and Magic

To summarize the core of my argument in this book, let us begin by reflecting on the notion of capitalism. It is generally applied to a particular economic system inaugurated in Europe no earlier than the sixteenth century (Wallerstein 1974–1989), but some theorists have dismissed such a historical discontinuity and instead attempted to trace processes of global capital accumulation several millennia back in time (Frank and Gills 1993).[1] In this latter view, the real modern discontinuity was the shift to fossil fuels in late eighteenth-century Britain (cf. Pomeranz 2000). From this perspective, we can understand the deliberations of classical political economists such as Ricardo and Marx as reflections not on a completely new mode of production, but on the new kind of society generated by steam power. Capital accumulation had been pervasive in stratified societies for millennia preceding the Industrial Revolution, but steam-driven technologies were the particular form of capital analyzed by Marx. Preindustrial forms of capital included farmland, livestock, roads, canals, armies, ships, and architecture. They, too, were material infrastructures that could be accumulated through the appropriation of labor-power and natural resources, and whose expansion in turn contributed to further such appropriation. This is the cross-cultural essence of capitalist power: a recursive relation between some kind of material infrastructure, on the one hand, and the capacity to make claims on other people's labor and resources, on the other.

This definition of capital, as based on appropriation, prompts us to rethink our concept of technology. Technology here refers not to the blueprints or engineering knowledge required to construct a particular machine or infrastructure, but to that machine or infrastructure as a material entity, which requires continuous inputs of fuel and maintenance work to function over time. Much of this book has been devoted to demonstrating that the continued operation of a given technology in this sense is contingent on asymmetric flows of energy, labor time, and/or other resources. Modern technology, in other words, is inextricably dependent on the rates of resource flows organized by the economy. This argument requires perspectives from both social and natural sciences. It thus may not make much sense to people trained in mainstream, neoclassical economics, who very rarely apply natural-science perspectives to their study of markets, but will probably be more congenial to heterodox economists with a background in ecological or Marxian economics.

It is appropriate to illustrate the dependence of technology on asymmetric resource flows by considering the emergence of steam power in late eighteenth-century Britain. The material metabolism of steam-driven textile factories in early industrial Britain hinged on the inexpensive labor employed in the American cotton fields and British coal mines, and on the inexpensive land available for the American cotton plantations. The world market prices of raw cotton versus cotton textiles confirm that a British factory owner who sold cotton cloth and bought raw cotton for the same sum of money in 1850 made a net gain in terms of invested labor time and more dramatically in terms of utilized space. Technological progress can thus be reconceptualized as the saving or liberation of human time and natural space in core regions of the world-system at the expense of time and space lost in the periphery. I have called this time–space appropriation (Hornborg 2006, 2013). A conventional economic analysis would only discern the flows of money, but by considering biophysical metrics such as embodied labor and land we can identify asymmetric flows of resources obscured by the apparent reciprocity of market prices. Asymmetric flows of embodied labor time in modern economies have been revealed by economists working in the Marxian tradition (e.g., Emmanuel 1972; cf. Simas et al. 2015), whereas the asymmetric flows of embodied land indicate that there is also what I have called an *ecologically* unequal exchange (Hornborg 1998, 2013; Dorninger and Hornborg 2015). The factor of production referred to as land can be subdivided into raw materials, energy, and eco-productive space. Recent research has shown that the core regions of the modern world-system—the United States, the European Union, and Japan—are all net importers of both embodied raw materials and embodied energy (Lenzen et al. 2012, 2013) as well as embodied space (Yu et al. 2013).

If the growth of industrial infrastructure in nineteenth-century Britain and modern core regions is geared to objectively measurable asymmetric flows of material resources, we need to ask a number of questions about the implications. Are these asymmetric flows morally reprehensible? Do they imply that technological progress in the United States, Europe, and Japan occurs at the expense of other parts of the world? If so, how has the science of economics managed to obscure such material asymmetries?

The position taken here is that the asymmetric flows of resources in the modern world-system are indeed morally reprehensible, because they imply that economic growth and technological progress in core regions largely occur at the expense of their trading partners in other parts of the world, and that levels of affluence in the cores are not possible to achieve universally. The net gains in embodied labor and land that for the past two centuries have been prerequisite to core expansion have implied a net loss of such resources for other parts of the world-system. The science of economics has obscured such material asymmetries and their moral implications by simultaneously excluding concerns with the material substance of traded commodities and concerns with morality. This view of world trade as neither material nor moral was established with the marginalist revolution in economics in the last decades of the nineteenth century. The neoclassical school of economics, which in this period dismissed the earlier concerns of political economy with embodied labor and land, continues to dominate the discipline—and economic policies worldwide—to this day. Considering how its preoccupation with market equilibrium systematically obscures the sources and mechanisms of global power and inequalities, it is a paradigmatic illustration of an ideology.

We have discussed how the discipline of economics simultaneously detached itself from concerns with materiality and morality. This detachment is epitomized by the meaninglessness, in the eyes of neoclassical economists, of the notion of unequal exchange. Although the concept is sometimes used to denote market power, such as monopoly, the theoretical framework of modern mainstream economics does not conceive of the occurrence of unequal exchange as defined here, that is, as an asymmetric flow of material resources contributing to growing inequalities. Neoclassical economic theory is exclusively concerned with market prices and monetary metrics. Unless transactors have exerted power over the market to set prices at other levels than would have been the outcome of free-market mechanisms, market transactions are by definition fair—or at least morally neutral. The mutual agreement signified by a given exchange rate obviates any additional concerns with reciprocity. This premise of modern economics is at odds with the outlook of political economy up until the marginalist revolution. Different schools of economic thought had previously emphasized different material aspects of commodity trade—whether the mercantilists' focus on precious metals, the Physiocrats' preoccupation with embodied land, or the classical economists' concern with embodied labor—but they all based their moral evaluations of trade on the material substance of the commodities exchanged. In abandoning such concerns with the substance of trade in favor of an exclusive focus on market equilibrium, neoclassical economics was left with no other criterion for morally evaluating trade than the extent to which it gave market mechanisms free reign. In this way, the simultaneous abandonment of material and moral assessments of specific commodity flows were concomitant. The incontrovertibly asymmetric and thus troubling global flows of embodied labor, land, energy, and materials remain conveniently outside mainstream economists' field of vision.

It is no coincidence that this mode of thinking about trade emerged in Britain at the apex of its colonial power, as the ideology accompanying the integration of the modern economic world-system. The material asymmetries of the Victorian world order are largely reproduced and intensified today, as illustrated by satellite images of nighttime lights revealing the contrast between the concentration of luminous technological infrastructure in Europe and North America, on the one hand, and the darkness of extractivist zones in Africa and South America, on the other. It is during one-and-a-half centuries of neoclassical economics that global inequalities have been dramatically aggravated, and yet we are told that its free-market agenda is designed precisely to liberate the global masses from poverty.

Economics must necessarily deal with morality. There can be no pretense that the rates at which humans exchange their labor and other resources on the market are automatically liberated from moral concerns. In order to grasp how the economists' preoccupation with money has entailed an illusory delegation of moral regulation to the mindless cybernetics of the market, we need to reflect on the very idea of money. The capacity to use money tokens to represent exchange relations, and to anchor expanding social structures to such extrasomatic artifacts, is uniquely human (cf. Strum and Latour 1987; Deacon 1997). For millennia, different kinds of money tokens have been used to concretize and regulate various transmutations of social reciprocity and indebtedness (Graeber 2011a). Like other sign systems, however, the management of money tokens has tended to become a game of its own, with rules continuously rewritten. The human makers of money have become subservient to the logic of their artifacts. This reification of interpersonal relations—and the concomitant inversion of power between human subjects and their money objects—has been deplored for two-and-a-half millennia (Bloch and Parry 1989). Aristotle called it chrematistics. St. Paul asserted that the love of money was the root of all evil. Thomas Aquinas proclaimed that greed was a cardinal sin. Marx aptly coined the concept of money fetishism. Meanwhile, however, another strain of thought progressively embraced the reification of human exchange. In the seventeenth and eighteenth centuries, philosophers such as Mandeville asserted that commerce was preferable to passions, paving the way for Adam Smith's *The Wealth of Nations* (Dumont 1977; Hirschman 1977). Stripped of its concern with labor value, Smith's celebration of market exchange remains foundational for modern economics.

This growing appreciation of commerce and accumulation since the early modern period reflects the increasing significance of money in the European merchant states of the time. In contrast to the largely agrarian societies of medieval Europe, the Genoese, Dutch, and British trading empires thrived primarily through the accumulation of money profits (Braudel 1992 [1979]; Arrighi 1994). Whereas agrarian societies were predominantly focused on harvesting solar energy through crops, livestock, and the labor of humans and draft animals, mercantile societies shifted focus to the management of money. Although modern concepts of energy did not appear until the mid-nineteenth century, humans have no doubt always been intuitively aware of the vital significance of the sun and what we know as photosynthesis. The growing preoccupation with money did not represent a liberation from these vital flows of energy, but a new strategy for gaining access to them—through purchasing power. The money accumulating in Europe provided access to vast solar-derived resources on other continents, including farmland, forests, livestock, game, fish, and not least human labor. Money can be viewed as a kind of fictive energy, in the sense that it is imagined as a vital flow that nourishes society.

It was in the context of such merchant capitalism that textile manufacturers in Britain in the late eighteenth century adopted steam power. In the struggle to increase reliable outputs of inexpensive cotton cloth for the world market, early British industrialists finally found fossil energy superior to water power (Malm 2016). Engineering was a necessary condition for this development, but it was not a sufficient one. The technological breakthrough represented by James Watt's steam engine would neither have occurred nor found to be useful if there had not been great global demand for inexpensive cotton cloth among West African slave traders and American slave owners (Inikori 1989, 2002). Slavery, slave plantations, and the triangular trade among Europe, Africa, and America were the foundations for the Industrial Revolution in several ways: by creating demand for its products, by providing cheap labor for harvesting its raw materials, and by offering the plantation as a template for the organization of industrial production. Industrial technology was thus contingent on processes in the eighteenth-century world-system.

However, to reconceptualize industrial machinery and infrastructure as crystallizations of social exchange relations is alien to modern thought. It is difficult for most modern people to equate bounded material objects with the intangible fields of relations which make them possible. Yet, this is what the science of ecology has taught us to do regarding biological organisms. The challenge is to apply the same perspective to technology. A functioning machine is no less dependent on continued resource flows than an organism.

As the resource flows provisioning a machine are contingent on market exchange rates, a further conclusion is that the existence of a technological infrastructure is contingent on the relative market prices of labor and other resources. To locally replace the expenditure of labor time with technologies requiring inputs of imported resources embodying labor time expended somewhere else is feasible only when wage differences between the two areas make it economically rational to do so. Globalized technologies, in other words, are products of arbitrage. This is the logic, for instance, behind contemporary exports to Europe and North America of various kinds of electrical equipment manufactured in China. Similar ratios between differently priced resources apply to the utilization of space. After Britain repealed its protectionist Corn Laws in 1846, the increasing imports of grain to England reflected lower production costs—including land rents and wages—in Prussia and North America, compared to those of England. To assert that the relatively low wages and land rents of nineteenth-century Prussia and twenty-first-century China offer these countries a comparative advantage in world trade, as would any follower of Ricardo, is a rather cynical euphemism for other countries taking advantage of their poverty and relatively lax environmental legislation. Using this terminology, slavery could be said to have granted the American colonies a comparative advantage in the form of low labor costs. Since the days of Ricardo, the language of free trade has thus promoted the displacement of both work and environmental loads to less affluent sectors of the world-system. It continues to justify increasing polarization and deepening inequalities. The mainstream denunciation of protectionism implies a dismissal of any policy to encourage self-sufficiency, reduce vulnerability, and restrain global transports. Given its implicit endorsement of asymmetric resource flows and economic polarization, the concept of globalization ultimately represents a less offensive way of talking about imperialism. In the context of current concerns with climate change, we should add that this same neoliberal worldview simultaneously promotes some of the most important sources of greenhouse gas emissions, such as industrial agriculture, deforestation, and global transport of bulk goods.

The nineteenth-century British shift to fossil fuels as a source of mechanical energy fundamentally transformed the conditions of economic rationality. It relaxed the ancient imperative to extract energy—for instance, firewood and fodder for draft animals—from the surface of the landscape and generally reduced the significance attributed to land as a factor of production. Ricardo concluded that the three factors of production—land, labor, and capital—were substitutable, so that, for example, a British shortage of land could be compensated for by an abundance of labor and capital. This observation was based on the experience of the Industrial Revolution, but it did not raise any concerns regarding the global implications of locally substituting capital for land. As the repeal of the Corn Laws illustrated, the appropriation, through trade, of the products of other nations' land is tantamount to environmental load displacement. The ecological relief which fossil fuels and imports of grain, timber, cotton, sugar, and other colonial produce granted to Britain represented a geographical space several times the total land area of Great Britain (Pomeranz 2000). Industrialization can thus be seen as a strategy for appropriating eco-productive space elsewhere in the world-system, and modern calculations of ecological footprints confirm that the same strategy has continued to be fundamental to economic growth and technological progress in core countries for over 200 years.

The marginalist revolution, which established the current hegemony of the neoclassical school in economics, reflects the limited capacity of earlier schools such as classical political economy to account for Britain's economic growth in the nineteenth century. Given the liberalization of trade and the net imports of both embodied land and labor, the accumulation of capital in England could obviously not be attributed to mercantilist trade policies, the fecundity of its soils, or the toil of its population. All such previous recipes for growth had been geared to theories of economic value based on the substance of traded commodities, whether concerned with precious metals, land fertility, or invested labor. Neoclassical economics abandoned all such considerations in favor of an exclusive concern with abstract exchange value, or "utility." Although the concept of utility in late nineteenth-century economics may have been inspired by the concept of energy in physics, and vice versa (Mirowski 1989), it paradoxically represents a definitive dismissal of material factors in the determination of market prices. With the marginalist revolution, mainstream economics rejected the ambition to derive the accumulation of money from the aggrandizement of some particular material metric. From now on, it was a science of pure chrematistics, detached from material considerations such as soil quality or inputs of labor time.

This mainstream detachment from material considerations is precisely what the heterodox schools of ecological and Marxian economics tend to criticize. Both schools have proposed that the exchange values which preoccupy neoclassical economists do not do justice to the intrinsic values of commodities exchanged on the market. Marxian economists phrase this discrepancy in terms of the difference between exchange value and use value, where the former is the market price of goods and services and the latter is conceived as their real material properties, such as the underpaid productive potential of labor-power. Many ecological economists similarly refer to underpaid natural values conceived, for instance, as ecosystem services or embodied energy. This convergence has generated similar approaches to the issue of unequal exchange, which in both schools has been theorized in terms of underpayment of real material values (Lonergan 1988; Foster and Holleman 2014).

At first sight, such a definition of unequal exchange may seem indistinguishable from the identification of asymmetric resource flows mentioned above, but the difference is important. It is theoretically untenable to posit the existence of material "values" that are not recognized by market actors. Economic values are determined by humans based on their preferences and assets. The material resources that are asymmetrically exchanged on the market should be conceptualized precisely as material resources, rather than as values. Regardless of whether it derives from human labor, draft animals, or fossil fuels, energy is not a value. Notions of underpaid material "use values" and "natural values" confuse physics and economics (Hornborg 2015). The inclination of ecological economics to equate energy and value has been as misleading as the Marxian distinction between them has been ambiguous and contradictory. It is obvious that the struggles of Marxian and ecological economics to conceptualize how market exchange orchestrates asymmetric material transfers are based on a valid intuition about the interaction of money and energy, but the assertion that labor-power and other forms of energy represent underpaid values is analytically flawed. The significance of asymmetric transfers of material resources is not that they represent underpaid values, but that they contribute to the physical expansion of productive infrastructure at the receiving end. The accumulation of such technological infrastructure may yield an expanding output of economic value, but this is not equivalent to saying that the resources that are embodied in infrastructure have an objective value in excess of their price. This objection to the conceptualization of unequal exchange in Marxian and ecological economics is necessary in order to offer an analytically rigorous, transdisciplinary argument with a solid grounding in both physics and economics.

An implication of this understanding of technology as contingent on ecologically unequal exchange is that we have reasons to be skeptical toward proposals for solving problems of sustainability that are founded on expectations of technological progress. Technological utopianism is an integral part of the modern worldview that accompanied the Industrial Revolution. To address the root of sustainability problems, we must instead acknowledge the destructive consequences of modern money. Money is what has made unequal market exchange, impoverishment, and technological overdevelopment possible to begin with. In making everything that humans desire commensurable and interchangeable, money automatically encourages the exchange of industrial commodities for increasing quantities of the natural resources that were used to produce them. The world market, in other words, rewards an accelerating dissipation of resources by providing access to ever more resources to dissipate. The conundrum we need to address is what policy implications can be drawn from Nicholas Georgescu-Roegen's (1971) recognition that economic processes simultaneously increase utility and entropy, or, in other words, monetary profits and material disorder.

Once we grasp the systematic discrepancy between the accumulation of money and the dissipation of matter and energy, the immediate response tends to be to propose some way of counteracting the transdisciplinary logic of Georgescu-Roegen's observation by better aligning money and energy. Such intuitively justifiable proposals have challenged neoclassical economics since its inception (Martinez-Alier 1987). However, to tie tokens of human exchange relations to energy would be as futile as to peg them to a gold standard. Human sign systems and thermodynamics follow completely different trajectories. Economics cannot be reduced to physics. Rather than try to make money reflect material reality, we should aspire to make it safeguard everybody's material needs. It is theoretically conceivable to design a complementary currency so as to insulate localized flows of necessities—and the integrity and resilience of communities as well as ecosystems—from the globalized arenas of financial speculation. Whether our priority is to avoid global financial crises or catastrophic climate change, we shall have to fundamentally redesign the operation of money.

Money and technology are both artifacts which organize and buttress human social relations. Particularly since the eighteenth century, the management of money and technology has had dramatic ramifications for the structure of world society. It is undisputable that these two kinds of artifacts are the pivotal props of globalization, regardless of whether we emphasize the integrative or exploitative aspects of such processes. The mainstream modern understandings of money and technology are that they are thoroughly rational and morally neutral inventions that can be contrasted against the "primitive" economies and magical practices of premodern societies. A central argument in this book, most explicitly articulated in chapter 6, is that such a contrast between the modern and premodern is difficult to sustain.

Although there have been several approaches to the definition of magic, my point of departure is the way in which social agency is delegated to artifacts. Where the agency of artifacts is contingent only on their objective physical properties, we may refer to such artifacts as technology, but where it is contingent on the subjective beliefs and perceptions of humans, we are in the realm of magic. As discussed in chapter 3, financial crises have recurrently revealed that money belongs to this latter category of artifacts. Already in the mid-nineteenth century, Marx clearly expressed this insight through the concept of money fetishism. For most modern people, however, it is cognitively more difficult to rethink technology in these terms. Yet, viewed from a global and transdisciplinary perspective, even modern technology is contingent on the subjective beliefs and perceptions of humans. The highly unequal access to infrastructures powered by exosomatic energy is contingent on global social structures of exchange, and these structures of exchange hinge on the flows and uneven distribution of money. The flows of money are no less magical than the money tokens themselves, for they are no less contingent on the subjective perceptions of humans. Money and technology together constitute a global game in which most players remain unaware of the extent to which the rules are both arbitrary and mutable, and in which the stakes are the relative distribution of workloads and environmental burdens. Accumulating money and technology is tantamount to shifting work and environmental degradation onto others. Although modern aspirations for "economic growth" and "development" are couched in the seemingly neutral language of economics and engineering, they mystify, through global magic, the appropriation of energy in the form of human time and natural space.

The central aim of this book has thus been to reconceptualize technology by theorizing the relation between energy and money. Energy is here defined as the material capacity to conduct work, and money—which of course includes all finance—as social strategies of exchange. The magic of finance has recently reinvigorated the United States as a major producer and exporter of oil, even though we know that American oil now represents very low net energy. This phenomenon is connected to Martinez-Alier's observation that Engels's dismissal of Podolinsky's suggestion to Marx was a missed opportunity for integrating thermodynamics and Marxist political economy. Modern eco-Marxists such as Burkett and Foster seem to want to renew the dialogue by exploring the relation between energy and the Marxian concept of surplus value, but wisely without reducing this relation to an energy theory of value. Neither embodied energy nor embodied labor directly translates into economic value, but the appropriation of energy— whether in the form of human labor or resources of the land (time or space)—is fundamental to capital accumulation, and probably an important empirical foundation for Marx's analysis of fossil-fuel capitalism. To restate this reconceptualization of technology as *appropriation*, I refer to the simple syllogism I traced many years ago (Hornborg 1998: 132): If technology is a matter of access to energy, and access to energy is a matter of money, then technology is a matter of money. This conclusion leads on to the question of how to distinguish between technology and magic, as it amounts to the observation that the material agency of technological artifacts is contingent on social strategies of exchange, while this contingency—as in all magic—is concealed from view.

The currently widespread insight that the world economy, unless fundamentally reorganized, is destined for collapse is often accompanied by a belief that the root of all evil is a specific "mode of production" that appeared in Europe in the sixteenth century and that can be overturned through revolution and replaced by a new, benevolent, and sustainable mode of production based on collective ownership and the production for use rather than for exchange. As history unfolds, this new mode of production, envisaged in the nineteenth century, increasingly assumes the appearance of a Utopia no less naïve than the millenarian movements in Melanesia mentioned at the outset of this book. The logic of modern capital is nothing less than the logic of money.[2] Rather than project our anxieties and indignation onto a reified, abstract but demonic "-ism," the collapse of which would give birth to a just and sustainable world society, we need to identify and transform the ideas (or memes, if you will) which continue to aggravate global inequalities and degradation. It is at the level of these seemingly self-evident cultural ideas that pervade our everyday lives—ultimately the notion of general-purpose money—that we can collectively decide to rewrite the rules of the game.

Capitalism and consumerism are *not* biological properties of our species. They are consequences of a specific human sign system which in a geological instant has come to dominate not only our world-encompassing society but also the biophysical world that it encompasses. Money is the idea that anything can be exchanged for anything else. It is an idea conventionally embodied in little pieces of metal or paper and more recently also in electronic digits on computer screens. If Gaia is afflicted by a virus, as has been metaphorically suggested, it is not humanity, but *money*. Humans have lived on this planet for a very long time without such money, and we can do it again.

In making everything interchangeable, money is the very foundation of the social condition we know as modernity. At the same time, it is the root of most of our ecological worries. The notion that rainforests are interchangeable with Coca-Cola goes against the grain of the complex hierarchy of levels of integration which life has been consolidating over the course of several billion years of evolution. Money thus jeopardizes the basic principles of life. It makes things commensurable that belong to very different levels of scale in time and space, and rewards—by providing access to ever more resources to dissipate—an accelerating dissipation of resources. The universally coveted phenomenon of "economic growth" inexorably implies an accelerating production of entropy.

As already Aristotle understood, money can serve as a useful tool for humans who want to exchange goods and services among themselves, but it can also become an end in itself, becoming our master instead of our servant. The critique of money fetishism has a 2,300-year-old genealogy from

Aristotle through St. Paul, Thomas Aquinas, Karl Marx, and Robert Owen, to name a few. Money is a transmutation and inversion of the Sacred, both signifying encompassment and abstraction, but while the Sacred is irreducible, for money nothing is sacred and everything reducible. Over the course of two-and-a-half millennia, money has been transformed from gold coins through paper bills to electrons, but its social and ecological implications continue to be governed by the pernicious logic of generalized interchangeability. Socially, most of the favors which humans do for one another have become commodities and services, the market prices of which are subject to meticulous calculation by both economists and people in general. The mathematics of the market have made our ancient concern with reciprocity superfluous, and it has become morally more important that pricing mechanisms operate freely than that the substance of commodity flows implies an equitable exchange of embodied labor time, land, energy, and materials. Ecologically, it has become increasingly evident that precisely those practices that are rewarded on the market simultaneously threaten, within a couple of generations, to make the Earth uninhabitable for human life.

Certainly, it is our biological capacity to produce abstract signs which has made money possible, but this does not mean that money in its present form is an inevitable expression of our biology. Our biological specificity also makes us capable of realizing what money does to the world, and to change the rules of the game so that our societies and ecosystems can survive even in the long run. It is perfectly possible, in principle, to design a money system which *strengthens* community and identity and simultaneously *reduces* vulnerability and environmental degradation.

So what is stopping us? Power, we say. The predictable intention of corporations and politicians to safeguard business as usual. Ways of thinking that will not change. But ultimately our economic and technological fixes derive from the peculiar inclination of humans to anchor our social relations in symbols and artifacts beyond our own bodies. As Strum and Latour clarified, this is what makes us different from all other animals. In various ways we delegate the patterned trajectories of our relations to the logic of external signs and artifacts. We let the interaction of *things* determine the future of our *social* interaction. When premodern Melanesians exchange prestigious *kula* shell ornaments with each other between the various islands, attention is focused on the features and histories of the ornaments themselves, rather than on the relations between the givers and recipients of gifts. We externalize our relations. A game of chess or a debate hinges on how pieces are moved or words are combined. Similarly, the logic of money is about the properties of money, not people, even though it determines human destinies. The world economy operates like a gigantic board game, in which the dice determines where we wind up. At the same time, as argued in chapter 6, the seemingly objective and relentless material agency of the artifacts to which we have delegated our destinies—even the power of oil—is contingent on human perceptions and social strategies.

I would like to add some reflections here about viewing the world economy as a game. It is conventionally represented as a "free" global market, but, as in all games, market "freedom" means the freedom to *obey the rules*, neither more nor less. However, compared with chess or popular board games, this game has some features and asymmetries which a chess player would hardly tolerate. First, the rules of the game have been withheld from the majority of the players during much of its history. Second, the minority who know the rules have reserved for themselves the right to change them over the course of the game. Third, it is not permitted to abandon the game. And fourth, in this game you can lose *everything*. It is literally a game of life and death.

It appears, however, that the Actor-Network theorists want us to embrace our fetishism. To attribute autonomous agency to things is for them a question of democracy beyond the human. In their view, saving Gaia from the disastrous scenarios of the Anthropocene requires attentiveness to the interests and purposes even of abiotic components of the planet, such as geological formations and artifacts. The recommendation is to abandon the humanist belief that there is something special about people. But my objection is that humanism is not necessarily equivalent to insensitivity to nonhuman life. To identify, as a "posthumanist," with matter in general—as if humans and rocks were commensurable—would be regressive and hardly of benefit to the biosphere. To stand a chance of avoiding ecological disaster we must instead mobilize and empower our uniquely human sensibility. Humanity still holds the potential of becoming an asset for the biosphere, rather than its most destructive component. As Rappaport (1994) put it, *we are that part of the biosphere which can reflect over itself.* This is an awe-inspiring perspective. We humans are that part of the Earth which can think and experience abstract care, which means that, potentially, we are able to envisage a more just and sustainable future.

Notes

2 Land, Energy, and Value in the Technocene

1. The Second Law of Thermodynamics states that entropy (disorder) will inevitably increase in an isolated system. As the Earth is not an isolated system, this did not pose a problem for the biosphere until humans began relying on finite deposits of fossil fuels.
2. The data compiled by Simas et al. (2015) allow us to conclude that the per capita import of embodied labor to the United States in 2007 was seven times greater than the per capita export (Dorninger and Hornborg 2015). Given a net import of over 0.3 person-year-equivalents per capita, even a one-child American household with an average living standard consumed a year of embodied labor from abroad, which is tantamount to saying that it had an invisible, full-time servant outside the nation's borders. A four-child household had two. Such figures corroborate the proposition that, to a large extent, modern technologies are not a replacement, but a *displacement*, of work.
3. The data presented in these sources tell us, for instance, that the consumption pattern of an average citizen of Japan in 2007 required net imports of almost 23 tons of embodied materials, over 30 gigajoules of embodied energy, and the products of almost a hectare of foreign land. These net imports all exemplify the phenomenon of environmental load displacement.
4. The notion of unpaid "costs" is, however, misleading, as it is theoretically unthinkable that the social and ecological disorder resulting from modern production processes could be neutralized by means of payments. Money cannot compensate for entropy. Even if it could, the internalization of "costs" would leave capitalists without profits and thus without incentives to produce.
5. The field of political ecology, in particular, has long struggled to reconcile the constructivist approaches predominant in anthropology and human geography, on the one hand, with objectivist approaches to biophysical nature, on the other (Escobar 1999). Adopting a much longer time perspective, environmental historians have traced our acknowledgment of revolutionary human–environmental interfusion to the late eighteenth century (Locher and Fressoz 2012). Humans have interfered with natural cycles for millennia (cf. Redman 1999), but the scale of interference following the Industrial Revolution is decisively transforming the biosphere.

6. For paradigmatic illustrations of world-system analysis, see Wallerstein (1974–1989) and Frank and Gills (1993). For edited collections attempting to integrate world-system analysis and global environmental change, see Goldfrank et al. (1999), Hornborg and Crumley (2007), and Hornborg et al. (2007).
7. Other alternatives to the Anthropocene include the Econocene (Norgaard 2013) and the Capitalocene (Malm and Hornborg 2014). The concept of the Capitalocene was coined by Andreas Malm at a seminar in Lund in 2009. It usefully emphasizes the role of capitalism in generating transformations of the biosphere, but might raise the objection that various forms of capital accumulation had caused ecological degradation, albeit at a lesser scale, for millennia before the Industrial Revolution (cf. Frank and Gills 1993; Redman 1999).
8. Latour's efforts to deconstruct distinctions between subject and object and between culture and nature apparently duplicate those of early German Romantics such as Friedrich Schelling in the period 1797–1806, and they raise the same philosophical objections (Wilding 2010).
9. Among the many costly, resource-intensive, and thus inherently privileged technologies that have been advocated as strategies to reduce (local) carbon dioxide emissions are nuclear power, carbon capture and storage, and photovoltaic energy (see chapter 7).
10. In this context of discussing anthropogenic environmental change, I see no reason to distinguish between industrial capitalism and the industrialism of purportedly noncapitalist societies such as the Soviet Union or China.

3 The Magic of Money

1. It thus seems as futile to deliberate on a general anthropological theory of value (Graeber 2001) as on a general anthropological theory of meaning. Anthropologists, sociologists, economists, linguists, and semioticians can provide us with tools for analyzing the *communication* of values and meanings in the practice of everyday life, but the idiosyncratic *sources* of these values and meanings are simply not amenable to theoretical analysis. Thousands of ethnographies describe how various attributions of value are communicated, but none can account for the specifics of the attributions themselves. Like other symbolic codes underlying cultural conventions, always arbitrary at the level of analysis (whether generating ritual, speech, or patterns of consumption), value can neither be reduced to theoretical constructs such as "socially necessary labor time" nor meaningfully accounted for by vacuous abstractions such as "utility" or "creative action."
2. Yet, Heilbroner assures us that, while mathematics "today pervades economics, formalizes it, and becomes its favored mode of expression, . . . no one actually confuses mathematics with economics" (1999 [1953]: 314). Georgescu-Roegen (1971) exposed the limitations of what he called arithmomorphism, which needs to be combined with a dialectical approach relying on words, instead of numbers (Mayumi 2009). According to Mayumi, "dialectical reasoning can be as correct as mathematical reasoning, but very often it can be even more penetrating" (ibid.: 1237).

3. Heilbroner (1999 [1953]: 109–115) does mention that the Utopian socialist Robert Owen in the early nineteenth century "naively" wanted to abolish money, but he never tells us why. There were, in fact, several movements to radically transform money in nineteenth-century England and the United States (North 2007: 41–61).
4. This dismay can be traced as far back as Aristotle, but few have expressed it as persuasively as Frederick Soddy (1926), a prominent ancestor figure for ecological economics (Daly 1996: 173–190).
5. Nicholas Georgescu-Roegen (1971) argued that the economic processes organized to enhance exchange values simultaneously entailed an inexorable physical dissipation of both matter and energy. One of his recurrent examples is the erosion of automobile tires into molecules of rubber randomly dissipated in the atmosphere. If his book had been written 30 or 40 years later, the most relevant example he could have chosen would have been the dissipation of carbon from fossil fuels, but then the focus of his observation would not have been on the impossibility of full recycling but on the limited sink capacity of the atmosphere. The dissipation of matter in economic processes, in other words, poses problems at both ends: the limited supply of resources as well as the disposal of waste and pollution. Although physicists have contested the claim that the laws of thermodynamics and the concept of entropy also apply to matter (cf. Ayres 1999; Mayumi 2009: 1243), they tend to agree with Georgescu-Roegen that complete recycling of matter is impossible. Regarding climate change, Georgescu-Roegen (1975: 358) had in 1975 recognized that "thermal pollution could prove to be a more crucial obstacle to growth than the finiteness of accessible resources" (cf. Mayumi 2009: 1248–1249).
6. We may ask to what extent the idiom of comparative advantage is in fact a euphemism for some nations taking advantage of other nations' lower labor (or other factor) costs, that is, work and/or environmental load displacement, whether applied to inexpensive wheat production in nineteenth-century Prussia or to the outsourcing of manufacture to twenty-first-century China.
7. By "neo-Physiocrat" ecological economics I mean those strands of this transdisciplinary field that, like the Physiocrats, view nature as the ultimate source of all economic value (see chapter 5).
8. The historical parallels and continuities between slavery and wage labor, popularly recognized in the notion of wage slavery, have been analyzed and discussed, for instance, by Graeber (2007) and McNally (2014). Although much effort has been devoted to establishing the difference between these two forms of commoditization of labor, epitomized in the ideal of freedom, the continuities are significant. As Graeber has remarked, for vast numbers of people over the past two centuries, the freedom of wage labor has been tantamount to an obligation to sell their freedom.
9. For an introduction to the formalist–substantivist controversy, see Wilk and Cliggett (2007).
10. Examples mentioned by Weatherford (1997) include the Banque Royale in 1720, the US Congress in 1780, the Bank of England in 1917, President Roosevelt in 1933, and President Nixon in 1971.

11. The agreement in Bretton Woods after the Second World War established the US dollar as the international reserve currency, backed by gold, obliging other nations to maintain ample supplies of dollars for purposes of trade (see Strange 1994; Eichengreen 2011).
12. Stiglitz (2010) notes the irony that it was the American politicians' policy of minimizing the role of government in the economy that ultimately, with the bailouts, gave it unprecedented control. In encouraging banks to become increasingly reckless, he argues, the bailouts have made "the problem of moral hazard . . . greater, by far, than it has ever been" (16–17).
13. Stiglitz is also explicitly critical of the self-serving behavior of financial institutions and the irresponsibility of the Federal Reserve. Rather than accuse bankers of excessive greed, however, he asks if we should really "blame the bankers for doing (perhaps a little bit better) what everyone in the market economy is supposed to be doing," that is, pursuing profit (Stiglitz 2010: 6).
14. Examples of "real" economic activities include the automobile industry and home construction (Foster and Magdoff 2009: 113).
15. Marxian economists deliberate on whether stagnation generates financialization, or vice versa (Foster and Magdoff 2009: 106). The causal relation between them is best understood as recursive.

5 Money as Fictive Energy: Unraveling the Relation between Economics and Physics

1. Much of the concern with net energy or EROI derives from Georgescu-Roegen's (1975) concept of accessible energy (Mayumi 2009: 1239).
2. The recognition that human labor, particularly as it was conceptualized in classical political economy, is a biophysical form of energy has been elaborated, for instance, by Rabinbach (1990) and in reflections on the parallels and continuities between slavery and the reliance on fossil fuels (Debeir et al. 1991 [1986]; Mouhot 2011; Nikiforuk 2012). The concern of ecological economists with declining net energy is no doubt related to the Marxian concern with declining profit rates due to an increasing ratio of machines to labor. Both recognize that the problem of diminishing returns increases with the growth of infrastructure, or technomass.
3. The obstacles to communication between Marxian and ecological economics can be illustrated by two revealing quotes, the first from Friedrich Engels in 1875 and the second from Robert Costanza in 1981. "No-one could convert specialized work into kilogrametres and determine salary differences based on that criteria" (Engels quoted in Martinez-Alier 1997: 231). "Can anyone seriously suggest that labor creates sunlight! The reverse is obviously more accurate" (Costanza quoted in Mirowski 1988: 817).
4. Although eco-Marxists such as Foster (2000) have attempted to repudiate the Promethean/modernist/urban bias of Marxian thought, it is nowhere more evident than in the casual reference, in the *Communist Manifesto*, to "the idiocy of

rural life." If the wastefulness and environmental destructiveness of modern capitalist agriculture is properly understood, the frugality and ecological wisdom of traditional peasant agriculture ought instead to be celebrated as a key to sustainability (cf. Martinez-Alier 1987, 1997; Mayumi 1991; Biel 2000; Greer 2008). The nineteenth-century Marxist derogation of rural life is a reflection of the same modernist/urban illusion that pervades the conviction in mainstream economics, since Ricardo, that human economies can liberate themselves from land (Mayumi 1991).

5. Even nineteenth-century capitalism had to reckon with costs for other forms of energy than labor (for instance, coal, water mills, draft animals), and there is no reason to suggest—or to propose that capitalists at the time believed—that investments in such inputs are less generative of value than those in labor. The attempt to incorporate ecology in the Marxian labor theory of value by discussing the "unpaid work" of nature is an analytically convoluted and misleading solution (cf. Odum 1996; Foster and Holleman 2014; Moore 2015). A common argument in defense of the Marxian labor theory of value is that it is merely an account of how capitalism actually operates, rather than a conviction about labor-power as uniquely generative of value. However, if the creation of surplus value is a specific feature of capitalism, rather than of labor in general, it seems contradictory to propose visions of socialism which outline alternative schemes for allocating this surplus.

6. Prominent illustrations of how the estimation of environmental externalities has become a preoccupation of some strands of ecological economics include calculations by Odum (1996) and Costanza et al. (1997). A similar confusion of physics and chrematistics is expressed in the persuasive but delusive notion of "ecological debt," which in principle *cannot* be repaid, only prevented from increasing.

7. Attempts to reconcile classical Marxist theory with ecological concerns tend to run into major analytical difficulties. A particularly instructive illustration is Jason Moore's (2015) aspiration to transcend the boundaries of Marxian dogma by suggesting that the exploitation of labor is founded on the appropriation of nature's unpaid "work/energy." Thus, for example, when the atmosphere serves as a sink for entropy in the form of greenhouse gases, it is "put to work as capital's unpaid garbage man" (101). Although fundamentally contradicting the definitions of central Marxian concepts such as labor, value, and proletarian (16–17)—conceding, among other things, that slaves are proletarians (99), fossil fuels and peasants are comparable resources (102), and nonhuman animals are "central to the production of surplus value" (93)—Moore generates 300 pages of diffuse and repetitive rhetoric in a convoluted struggle to cling, against all odds, to the labor theory of value. If, for Moore, labor-power is merely one of four "Cheap Natures" (in addition to food, energy, and raw materials), we are prompted to ask if not *all* of these resources are ultimately embodied *land*? Moore's attempts to theorize the appropriation of ecological resources within a Marxian framework yields a turgid and obscure idiom for observing that capitalists make money by keeping costs lower than income. If we acknowledge the crucial role of energy and other natural resources in the intensified production processes yielding economic value, it is unfortunate that some scholars feel a need to engage in such tortuous struggles to cling to a labor theory

of value. Moore's analysis is not enhanced by his awkward attempt to present it as a "post-Cartesian" project, nor by his ingenuous assertions that "entropy is reversible and cyclical" (97).
8. This strain of ecological economics reiterates the proposals of the so-called Technocrat movement to align economic categories with those of physics.
9. Of course there are minimum material (for instance, nutritional) requirements for human survival, without which the "enjoyment of life" would not be possible, but for the majority of modern humans they are irrelevant for the determination of what Burkett calls "use values."
10. Rather than attempt to patch up such shortcomings of classical Marxian theory, for instance, by advocating advanced technological solutions (Schwartzman 1996) or envisaging a high-tech society properly paying for the "work of nature" (Moore 2015), a grasp of the relation between economics and thermodynamics revealed by Georgescu-Roegen (1971) can only lead to the tradition of thought referred to as "Degrowth" and represented by thinkers such as André Gorz, Serge Latouche, and Ivan Illich (D'Alisa et al. 2015).
11. The paradigmatic case is how the "apparently equal exchange of the worker's labour-power for its value . . . 'turns into its opposite . . . the dispossession of his labour'" (Burkett 2005b: 192, reference to Karl Marx).
12. The incomprehensible implication is that labor-power invested in export production in commercial, precapitalist civilizations should have been less significant for the creation of profit than in nineteenth-century capitalism. However, the alleged uniqueness of the modern capitalist mode of production in generating global inequalities, environmental degradation, and money fetishism has been challenged from various directions in recent years (for instance, Frank and Gills 1993; Redman 1999; Graeber 2011a). The historical discontinuity represented by nineteenth-century capitalism was primarily the shift to fossil fuels as a source of mechanical energy (cf. Pomeranz 2000), and this shift was in itself contingent on global conjunctures such as the demand for cotton textiles in the lucrative triangular trade across the Atlantic.
13. If Podolinsky's reasoning seems rudimentary in comparison with Odum's, we should recall that it preceded Odum's work by almost a century. It is noteworthy that there is even a historical link between the Marxian thermodynamics of Podolinsky and those of Odum (Martinez-Alier 1987: 231, n. 17).

6 Agency, Ontology, and Global Magic

1. A fundamental constraint of the perspectivist approach, however, is that it will always remain confined to human representations of nonhuman perspectives. It will never be able to say anything specific about how nonhumans actually experience the world (cf. Descola 2014: 272).
2. Elder-Vass's (2015) critique of ANT exposes its self-contradictions with regard to anthropocentrism and the existence of phenomena beyond human discourse. Condemned by his own propositions to continuous self-reference, Latour's

deliberations can make no claim to account for the specific ways in which nonhuman physical forms assume the role of "actants" that is not constrained by his own particular vantage point. But to be compelled to include the human observer-participant in every attempt to represent something—whether human or nonhuman—is as unreasonable as it is unfeasible. To posit, where applicable, a recursive (i.e., mutually constitutive) *relation* between reference and referent—"the knower and the known" (cf. Hornborg 1996: 52)—is not equivalent to positing their *conflation*.
3. In a rejoinder, Latour (2014) clearly does not agree. There is a widespread failure in anthropology to distinguish between human claims that abiotic things—such as sacred mountains or mummified ancestors—are animate, on the one hand, and the issue if they are actually alive, on the other. To respect such claims as statements about human sentiments and relations should not be confused with skepticism vis-à-vis biological or semiotic definitions of life (cf. Kohn 2013, 2014). Assertions that a mountain is animate may be understood as an appropriate counterbalance to the equally fetishistic claim that a corporation is a *person* (Martin 2014: 107), but we should be aware that we are talking about human sentiments and relations rather than actually attributing personhood to geological formations. It will require profoundly humanist—rather than "posthumanist"—convictions in order to establish sensitive and sustainable relations to the remainder of the biosphere. This will entail enhancing our extraordinary human capacity for subjectivity, rather than regressively reducing ourselves to equivalents of rocks and tools.
4. The mechanical philosophy of the scientific revolution of seventeenth-century Europe demanded "certainty of demonstration by experiment and dissection" (Tambiah 1990: 21), which implied systematically confining attention to the visible internal operation of machines and organisms, excluding the invisible external networks of relations in which they are embedded. This detached, scientific understanding of the "truth" of technological efficacy is fundamentally different from the performative efficacy of magic (cf. Graeber 2001: 241–244), which operates on social relations and subjective experiences.
5. Significantly, the invention of the slot machine enabled even coins to assume technological functions, alongside the magic that Marx called money fetishism: when we buy a Coke or enter a public bathroom we might reflect over the fact that magical objects can be converted into technology.
6. Among the Matis, the reluctance to engage with artifacts manufactured by others is clearly related to their historical experience of epidemics introduced by outsiders (Erikson 2009: 179).
7. Whether we are prepared to acknowledge them or not, probably all of us do have such assumptions. For a convincing argument in this direction, see Heywood (2012: 146).
8. Chris Gregory (2014) notes that posthumanist attributions of agency or even intentionality to things are "from a Marxist perspective, a classic example of fetishistic thought of an animistic kind," and "quite literally a form of spiritualism" that no humanist can accept. "For the humanist," he continues, such "assumptions are part of the problem to be explained in the current era of hypercommodity fetishism, not a solution to it" (48, 62).

9. The potential dangers of eating insufficiently desubjectivized food are clarified in the distinction made by Carlos Fausto (2007)—building on the observations of Beth Conklin, Aparecida Vilaça, and others—between cannibalism, defined as the appropriation of subjective aspects of other humans, and eating "familiarized" human remains.
10. This does seem an appropriate ideology for neoliberal social science (cf. Gregory 2014). The ideological bottom line of ANT is to shift political responsibility from humans to things (Martin 2014: 105–107). Although offered as a joke, it is perhaps no coincidence that Latour (2005: 5) quotes Mrs. Thatcher's slogan, "There is no such thing as a society." A refreshing contrast is Kim Fortun's (2014) criticism of Latour's conceptual framework for disregarding the various environmental and medical externalizations that are inherent in "industrial logic." Although Fischer (2014) largely follows Latour into esoteric language games, he similarly challenges him to address "the widening inequalities and devastations of our current cannibal economies, consuming the lives of some for the luxury of others" (349). The widely recognized inability of ontological anthropology to deliver political critique (Bessire and Bond 2014; Vigh and Sausdal 2014: 63) raises concerns over its official prominence at recent meetings of the American Anthropological Association.
11. Note that capitalist violence can also be represented as highly intentional, as illustrated by Santos-Granero and Frederica Barclay (2011) in their article "Bundles, stampers, and flying gringos."
12. In a more recent paper, Santos-Granero (2013) shows that such notions of ensoulment are widespread also in Euro-American societies.
13. Rather than dismiss all subject–object distinctions as symptoms of a false consciousness foundational to modernity, we need to understand them as statements about relations of power (cf. Martin 2014: 111). Fetishism can be understood as the attribution of power—the displacement of responsibility—to objects within networks of social relations where the political agency of humans is not apparent.
14. This was convincingly argued by Alfred Gell (1992), not least in his chapter "The Technology of Enchantment and the Enchantment of Technology." Particularly interesting in this context is Gell's (1992: 62, n. 3) observation that modern technologies are cognate to magic in the sense that they tend to create illusions of costless production while displacing social and environmental "costs" elsewhere (cf. Fortun 2014).
15. Gregory's (2014: 57–60) review of the approach of so-called cultural economy reveals how *homo economicus* has been reborn as the market's calculative agencies, largely embodied in financial traders as "thinking subjects" pitting hope against uncertainty in their struggle to exploit differences in time, space, and human knowledge. A price, in this harmonious view of the market, is an "acceptable compromise" (ibid.). The market price of oil, no less than that of other commodities, hinges on human perceptions (Yergin 1991: chapters 34–35). This means that the seemingly material agency of oil is ultimately contingent on social relations and subjectivities.

16. Although the concept of globalization is indispensable in communicating this aspect of modern technology, we should keep in mind that it was introduced to replace the neo-Marxist concept of imperialism (Gregory 2014: 56–58). Rather than immersing ourselves in alternative ontologies and denying the reality of "a common world" (Goldman 2009: 113), anthropologists would do well to contemplate the incontrovertible material inequalities evident in global statistics on purchasing power and physically visible even in nightly satellite images of the world.
17. This conclusion demands a precise definition of the word "modern" relevant to the context. By "modern" I here mean an aspiration to emancipate the human majority by exposing and abandoning representations of society that are magical in the sense that they mystify power relations, not least by attributing social inequalities to the autonomous agency of nonliving objects.

7 The Political Ecology of Technological Utopianism

1. A revised version of this article was included in Hornborg (2013).
2. In comparison, the oil extracted in the beginning of the age of petroleum had a net energy of around 100 times the input.
3. Spain is often presented as one of the most ideal countries for developing solar power, but the annual Spanish subsidies for this energy source (around 2.3 billion euros) may have been a major contribution to the nation's financial crisis (Prieto and Hall 2013: 28, 36).

8 Redesigning Money to Curb Globalization and Increase Resilience

1. Although several different designations occur—for instance, local, community, or alternative currencies—the concept of complementary currency seems most precise for the proposal presented here, as it does not aspire to replace the normal currency with a geographically more restricted one, but to provide an option alongside it.
2. A convenient way of distinguishing the range of local goods would be to mark them as such, but such marking would of course vary between shops in different places. Rather than amount to a number of geographically distinct, local currencies, this system would mean *one* complementary currency for the whole nation, but with an inbuilt inclination to generate localized but overlapping circuits of exchange.
3. Of course, this is not to deny that some people might consider looking for other jobs, for instance, those who today profit from financial speculation or from industries such as the production of and international trade in foodstuffs or petroleum. As the rationale of this proposal is precisely to transform the current organization of the global economy, major long-term structural adjustments are to be expected, much as the world's population has learned to accept over the past two centuries.

4. Following the introduction of digital money, the proportion of foreign exchange transactions that pertain to speculation in currencies now dwarfs the insignificant percentage pertaining to the purchase and sale of real goods and services.
5. Most social scientists would reject the notion that social transformations can be understood as learning processes, and it is reassuring to find a forest ecologist similarly wondering "how does an ecosystem 'learn' and 'adapt'?" (Park 2011: 339).
6. However, it should be pointed out that Brand and Jax (2007) are generally critical of transdisciplinary uses of the concept of resilience, advocating instead a clearly specified, descriptive usage of the term within ecological science. They argue somewhat ambiguously that, while sharing the term may have facilitated communication across disciplinary boundaries, the understandings of its precise meaning would differ (ibid.: 9).
7. Cf. Zygmunt Bauman's (1998, 2011) observation that globalization for some has meant localization for others.
8. This is not to deny that a minority of speculators has profited enormously from these crises, but the problems of unrealistic standards of consumption and of increasing fiscal deficits are not solved simply by redistributing such profits for collective use.
9. It is thus perhaps fitting that indigenous Quechua women are portrayed on the cover of Berkes and Folke's (1998) classical volume *Linking Social and Ecological Systems*.

Conclusions: Money, Technology, and Magic

1. Export production of goods such as textiles and arms has occurred for millennia. Production for exchange was pervasive enough in ancient Greece to incite Aristotle to coin the distinction between use value and exchange value. The incentive to make monetary profits from such exchange has prompted successive rationalization of production processes of which the turn to steam power was merely a recent and spectacular example. The coercive power of chrematistics can be traced to ancient Mesopotamia, and contemporary "wage slavery" is a variation on a theme that goes back thousands of years (Graeber 2011a).
2. Although presenting insightful accounts of the global, social and ecological destructiveness of "capitalism," several recent contributions (e.g., Harvey 2010; Foster et al. 2010; Klein 2014; Moore 2015) unfortunately fail to achieve a critical, reflexive distance to the cultural phenomenon of money. Even for most Marxists, money is like water to fish. These authors seem to take the idea of general-purpose money for granted, never once mentioning it as a problem, yet the destructive logic of capital is contingent on the way we all endorse the use of money in our everyday lives. Moreover, as I hope to have shown in this book, if we fail to see the problem with money, we fail to see the problem with technology.

References

Adas, Michael. 1989. *Machines as the measure of men: Science, technology, and ideologies of Western dominance.* Ithaca: Cornell University Press.
Adas, Michael. 2006. *Dominance by design: Technological imperatives and America's civilizing mission.* Cambridge, MA: MIT Press.
Albritton Jonsson, Fredrik. 2014. The origins of cornucopianism: A preliminary genealogy. *Critical Historical Studies* 1(1):151–168.
Alsamawi, Ali, Joy Murray, and Manfred Lenzen. 2014. The employment footprint of nations: Uncovering master-servant relationships. *Journal of Industrial Ecology* 18(1):59–70.
Altman, Ida. 2011. The Spanish Atlantic, 1650–1780. In *The Oxford handbook of the Atlantic world 1450–1850*, edited by Nicholas Canny and Philip Morgan, 183–200. Oxford: Oxford University Press.
Altvater, Elmar. 1990. The foundations of life (nature) and the maintenance of life (work): The relation between ecology and economics in the crisis. *International Journal of Political Economy* Spring:10–34.
Amin, Samir. 1976. *Unequal development.* New York: Monthly Review Press.
Andersen, Otto. 2013. *Unintended consequences of renewable energy.* London: Springer.
Århem, Kaj. 1996. The cosmic food web: Human-nature relatedness in the northwest Amazon. In *Nature and society: Anthropological perspectives*, edited by Philippe Descola and Gísli Pálsson, 185–204. London: Routledge.
Arrighi, Giovanni. 1994. *The long twentieth century: Money, power and the origins of our times.* New York: Verso.
Ayres, Robert U. 1998. The second law, the fourth law, recycling and limits to growth. *INSEAD Working Paper* No. 98/38/EPS/CMER. Fontainebleau: INSEAD.
Ayres, Robert U. 1999. The second law, the fourth law, recycling and limits to growth. *Ecological Economics* 29:473–483.
Balée, William, and Clark L. Erickson, eds. 2006. *Time and complexity in historical ecology: Studies in the neotropical lowlands.* New York: Columbia University Press.
Basso, Ellen. 2011. Amazonian ritual communication in relation to multilingual social networks. In *Ethnicity in ancient Amazonia: Reconstructing past identities from archaeology, linguistics, and ethnohistory*, edited by Alf Hornborg and Jonathan D. Hill, 155–171. Boulder: University Press of Colorado.
Bateson, Gregory. 1972. *Steps to an ecology of mind.* Frogmore: Paladin.

Baudrillard, Jean. 1981 [1972]. *For a critique of the political economy of the sign.* St. Louis: Telos.
Bauman, Zygmunt. 1998. *Globalization: The human consequences.* Cambridge: Polity.
Bauman, Zygmunt. 2011. *Collateral damage: Social inequalities in a global age.* Cambridge: Polity.
Benton, Ted. 1989. Marxism and natural limits: An ecological critique and reconstruction. *New Left Review* 178:51–86.
Berkes, Fikret. 1999. *Sacred ecology: Traditional ecological knowledge and resource management.* Philadelphia: Taylor & Francis.
Berkes, Fikret, Johan Colding, and Carl Folke, eds. 2003. *Navigating social-ecological systems: Building resilience for complexity and change.* Cambridge: Cambridge University Press.
Berkes, Fikret, and Carl Folke, eds. 1998. *Linking social and ecological systems: Management practices and social mechanisms for building resilience.* Cambridge: Cambridge University Press.
Besley, Tim, and Peter Hennessy. 2009. The global financial crisis – Why didn't anybody notice? *British Academy Review* 14:8–10.
Bessire, Lucas, and David Bond. 2014. Ontological anthropology and the deferral of critique. *American Ethnologist* 41(3):440–456.
Biel, Robert. 2000. *The new imperialism: Crisis and contradictions in North/South relations.* London: Zed Books.
Biel, Robert. 2006. The interplay between social and environmental degradation in the development of the international political economy. *Journal of World-Systems Research* 12(1):109–147.
Biel, Robert. 2012. *The entropy of capitalism.* Chicago: Haymarket Books.
Biersack, Aletta, and James B. Greenberg, eds. 2006. *Reimagining political ecology.* Durham: Duke University Press.
Bird-David, Nurit. 1999. "Animism" revisited: Personhood, environment, and relational epistemology. *Current Anthropology* 40S:67–91.
Blaikie, Piers, and Harold Brookfield. 1987. *Land degradation and society.* London: Methuen.
Blanc, Jérôme, ed. 2012. Thirty years of community and complementary currencies: A review of impacts, potential and challenges. Special issue of *International Journal of Community Currency Research* 16.
Blaut, James M. 1993. *The colonizer's model of the world: Geographical diffusionism and Eurocentric history.* New York: The Guilford Press.
Blaut, James M. 2000. *Eight Eurocentric historians.* New York: The Guilford Press.
Bloch, Maurice, and Jonathan Parry. 1989. Introduction: Money and the morality of exchange. In *Money and the morality of exchange*, edited by Jonathan Parry and Maurice Bloch, 1–32. Cambridge: Cambridge University Press.
Blower, David. 2000. The many facets of *mullu*: More than just a *Spondylus* shell. *Andean Past* 6:209–228.
Bogadóttir, Ragnheidur. 2012. Fleece: Imperial metabolism in the Pre-Columbian Andes. In *Ecology and power: Struggles over land and material resources in the past,*

present and future, edited by Alf Hornborg, Brett Clark, and Kenneth Hermele, 83–96. London: Routledge.

Bohannan, Paul. 1955. Some principles of exchange and investment among the Tiv. *American Anthropologist* 57:60–70.

Boomert, Arie. 1987. Gifts of the Amazons: "green stone" pendants and beads as items of ceremonial exchange in Amazonia and the Caribbean. *Antropologica* 67:33–54.

Bourdieu, Pierre. 1984 [1979]. *Distinction: A social critique of the judgement of taste.* London: Routledge.

Brand, Fridolin Simon, and Kurt Jax. 2007. Focusing the meaning(s) of resilience: Resilience as a descriptive concept and a boundary object. *Ecology and Society* 12(1):23.

Braudel, Fernand. 1992 [1979]. *The perspective of the world. Civilization and capitalism, 15th–18th centuries*, vol. 3. Berkeley: University of California Press.

Brennan, Teresa. 2000. *Exhausting modernity: Grounds for a new economy.* London: Routledge.

Brennan, Teresa. 2003. *Globalization and its terrors: Daily life in the West.* London: Routledge.

Bryant, Raymond L., and Sinead Bailey. 1997. *Third World political ecology.* London: Routledge.

Bunker, Stephen G. 1985. *Underdeveloping the Amazon: Extraction, unequal exchange, and the failure of the modern state.* Chicago: The University of Chicago Press.

Burbank, Jane, and Frederick Cooper. 2010. *Empires in world history: Power and the politics of difference.* Princeton: Princeton University Press.

Burkett, Paul. 2003. The value problem in ecological economics: Lessons from the Physiocrats and Marx. *Organization & Environment* 16(2):137–167.

Burkett, Paul. 2005a. Entropy in ecological economics: A Marxist intervention. *Historical Materialism* 13(1):117–152.

Burkett, Paul. 2005b. *Marxism and ecological economics: Toward a red and green political economy.* Leiden: Brill.

Burkett, Paul, and John Bellamy Foster. 2006. Metabolism, energy, and entropy in Marx's critique of political economy: Beyond the Podolinsky myth. *Theory and Society* 35:109–156.

Carter, Benjamin P. 2011. *Spondylus* in South American prehistory. In *Spondylus in prehistory: New data and approaches. Contributions to the archaeology of shell technologies*, edited by Fotis Ifantidis and Marianna Nikolaidou, 63–89. Oxford: BAR International Series 2216.

Chakrabarty, Dipesh. 2009. The climate of history: Four theses. *Critical Inquiry* 35:197–222.

Chaplin, Joyce E. 2011. The British Atlantic. In *The Oxford handbook of the Atlantic world 1450–1850*, edited by Nicholas Canny and Philip Morgan, 219–234. Oxford: Oxford University Press.

Clark, Brett, and John Bellamy Foster. 2012. Guano: The global metabolic rift and the fertilizer trade. In *Ecology and power: Struggles over land and material resources*

in the past, present and future, edited by Alf Hornborg, Brett Clark, and Kenneth Hermele, 68–82. London: Routledge.
Costa, Luiz, and Carlos Fausto. 2010. The return of the animists: Recent studies of Amazonian ontologies. *Religion and Society: Advances in Research* 1(1):89–109.
Costanza, Robert. 1980. Embodied energy and economic evaluation. *Science* 210: 1219–1224.
Costanza, Robert, Ralph d'Arge, Rudolf de Groot, Stephen Farber, Monica Grasso, Bruce Hannon, Karin Limburg, Shahid Naeem, Robert V. O'Neill, Jose Paruelo, Robert G. Raskin, Paul Sutton, and Marjan van den Belt. 1997. The value of the world's ecosystem services and natural capital. *Nature* 387:253–260.
Croll, Elisabeth, and David Parkin, eds. 1992. *Bush Base – Forest Farm: Culture, environment and development*. London: Routledge.
Crosby, Alfred W. 1986. *Ecological imperialism: The biological expansion of Europe, 900–1900*. Cambridge: Cambridge University Press.
Crosby, Alfred W. 2006. *Children of the sun: A history of humanity's unappeasable appetite for energy*. New York: W. W. Norton.
Crump, Thomas. 1981. *The phenomenon of money*. London: Routledge & Kegan Paul.
Crutzen, Paul. 2002. Geology of mankind. *Nature* 3, January 23.
D'Alisa, Giacomo, Federico Demaria, and Giorgos Kallis, eds. 2015. *Degrowth: A vocabulary for a new era*. London: Routledge.
D'Altroy, Terence N. 2001. Politics, resources, and blood in the Inka Empire. In *Empires: Perspectives from archaeology and history*, edited by Susan E. Alcock, Terence N. D'Altroy, Kathleen D. Morrison, and Carla M. Sinopoli, 201–226. Cambridge: Cambridge University Press.
D'Altroy, Terence N. 2002. *The Incas*. Oxford: Blackwell.
Daly, Herman E. 1977. *Steady-state economics: The economics of biophysical equilibrium and moral growth*. San Francisco: W. H. Freeman.
Daly, Herman E. 1996. *Beyond growth*. Boston: Beacon Press.
Daly, Herman E. 2012. Nationalize money, not banks. http://www.paulcraigroberts.org/2012/07/30/nationalize-money-not-banks-herman-daly/ [accessed November 29, 2014].
Daniels, Farrington. 1964. *Direct use of the sun's energy*. New York: Ballantine Books.
Deacon, Terrence. 1997. *The symbolic species: The co-evolution of language and the human brain*. London: Penguin.
Debeir, Jean-Claude, Jean-Paul Deléage, and Daniel Hémery. 1991. *In the servitude of power: Energy and civilization through the ages*. London: Zed Books.
Delucchi, Mark A., and Mark Z. Jacobson. 2011. Providing all global energy with wind, water, and solar power, Part II: Reliability, system and transmission costs, and policies. *Energy Policy* 39:1170–1190.
Descola, Philippe. 1994. *In the society of nature: A native ecology in Amazonia*. Cambridge: Cambridge University Press.
Descola, Philippe. 1996. Constructing natures: Symbolic ecology and social practice. In *Nature and society: Anthropological perspectives*, edited by Philippe Descola and Gísli Pálsson, 82–102. London: Routledge.

Descola, Philippe. 2013. *Beyond nature and culture.* Chicago: The University of Chicago Press.
Descola, Philippe. 2014. All too human (still): A comment on Eduardo Kohn's *How Forests Think. HAU: Journal of Ethnographic Theory* 4(2):267–273.
Descola, Philippe, and Gísli Pálsson, eds. 1996. *Nature and society: Anthropological perspectives.* London: Routledge.
Diamond, Jared. 1997. *Guns, germs, and steel: The fates of human societies.* New York: W. W. Norton.
Diamond, Jared. 2005. *Collapse: How societies choose to fail or succeed.* New York: Viking.
Dittmer, Kristofer. 2013. Local currencies for purposive degrowth? A quality check of some proposals for changing money-as-usual. *Journal of Cleaner Production* 54:3–13.
Dittrich, Monika, and Stefan Bringezu. 2010. The physical dimension of international trade: Part I. Direct global flows between 1963 and 2005. *Ecological Economics* 69(9):1838–1847.
Dobson, Ross V. G. 1993. *Bringing the economy home from the market.* Montreal: Black Rose Books.
Dorninger, Christian, and Alf Hornborg. 2015. Can EEMRIO analyses establish the occurrence of ecologically unequal exchange? *Ecological Economics* 119:414–418.
Douthwaite, Richard. 1999. *The ecology of money.* Totnes: Green Books.
Duchin, Faye, and Stephen H. Levine. 2012. Embodied resource flows in a global economy: An approach for identifying the critical links. *Journal of Industrial Ecology* 17(1):65–78.
Duffield, Mark. 2008. *Development, security and unending war: Governing the world of peoples.* Cambridge: Polity.
Dumont, Louis. 1977. *From Mandeville to Marx: The genesis and triumph of economic ideology.* Chicago: The University of Chicago Press.
Ehrlich, Paul R., and John P. Holdren. 1971. Impact of population growth. *Science* 171:1212–1217.
Eichengreen, B. 2011. *Exorbitant privilege: The rise and fall of the dollar and the future of the international monetary system.* Oxford: Oxford University Press.
Elder-Vass, Dave. 2015. Disassembling Actor-Network Theory. *Philosophy of the Social Sciences* 45(1):100–121.
Emmanuel, Arghiri. 1972. *Unequal exchange: A study of the imperialism of trade.* New York: Monthly Review Press.
Erb, Karl-Heinz, Fridolin Krausmann, Wolfgang Lucht, and Helmut Haberl. 2009. Embodied HANPP: Mapping the spatial disconnect between global biomass production and consumption. *Ecological Economics* 69:328–334.
Erikson, Philippe. 2009. Obedient things: Reflections on the Matis theory of materiality. In *The occult life of things: Native Amazonian theories of materiality and personhood,* edited by Fernando Santos-Granero, 173–191. Tucson: The University of Arizona Press.
Escobar, Arturo. 1999. After nature: Steps to an anti-essentialist political ecology. *Current Anthropology* 40(1):1–30.
Fausto, Carlos. 2007. Feasting on people: Eating animals and humans in Amazonia. *Current Anthropology* 48(4):497–530.

Financial Crisis Inquiry Commission. 2011. *The financial crisis inquiry report.* New York: Public Affairs.
Fischer, Michael M. J. 2014. The lightness of existence and the origami of "French" anthropology: Latour, Descola, Viveiros de Castro, Meillassoux, and their so-called ontological turn. *HAU: Journal of Ethnographic Theory* 4(1):331–355.
Fischer-Kowalski, Marina, and Christof Amann. 2001. Beyond IPAT and Kuznets curves: Globalization as a vital factor in analyzing the environmental impact of socio-economic metabolism. *Population and Environment* 23(1):7–47.
Folke, Carl. 2006. Resilience: The emergence of a perspective for social-ecological systems analyses. *Global Environmental Change* 16:253–267.
Fortun, Kim. 2014. From Latour to late industrialism. *HAU: Journal of Ethnographic Theory* 4(1):309–329.
Foster, George. 1965. Peasant society and the image of limited good. *American Anthropologist* 67(2):293–315.
Foster, John Bellamy. 2000. *Marx's ecology: Materialism and nature.* New York: Monthly Review Press.
Foster, John Bellamy. 2008. Peak oil and energy imperialism. *Monthly Review* 60(3):12–33.
Foster, John Bellamy. 2013. The epochal crisis. *Monthly Review* 65(5).
Foster, John Bellamy, and Paul Burkett. 2004. Ecological economics and classical Marxism: The "Podolinsky business" reconsidered. *Organization & Environment* 17(1):32–60.
Foster, John Bellamy, and Paul Burkett. 2008. Classical Marxism and the Second Law of Thermodynamics: Marx/Engels, the heat death of the universe hypothesis, and the origins of ecological economics. *Organization & Environment* 21(1):3–37.
Foster, John Bellamy, and Brett Clark. 2004. Ecological imperialism: The curse of capitalism. *Socialist Register* 40:186–201.
Foster, John Bellamy, Brett Clark, and Richard York. 2010. *The ecological rift: Capitalism's war on the Earth.* New York: Monthly Review Press.
Foster, John Bellamy, and Hannah Holleman. 2014. The theory of unequal ecological exchange: A Marx-Odum dialectic. *The Journal of Peasant Studies* 41(2):199–233.
Foster, John Bellamy, and Fred Magdoff. 2009. *The great financial crisis: Causes and consequences.* New York: Monthly Review Press.
Frank, Andre Gunder. 1967. *Capitalism and underdevelopment in Latin America.* New York: Monthly Review Press.
Frank, Andre Gunder. 1998. *ReOrient: Global economy in the Asian age.* Berkeley: University of California Press.
Frank, Andre Gunder. 2007. Entropy generation and displacement: The nineteenth-century multilateral network of world trade. In *The world system and the Earth system: Global socioenvironmental change and sustainability since the Neolithic,* edited by Alf Hornborg and Carole Crumley, 303–316. Walnut Creek: Left Coast Press.
Frank, Andre Gunder, and Barry K. Gills, eds. 1993. *The world system: Five hundred years or five thousand?* London: Routledge.

Friedman, Jonathan, and Kajsa Ekholm Friedman. 2013. Globalization as a discourse of hegemonic crisis: A global systemic analysis. *American Ethnologist* 40(2):244–257.
Garnsey, Peter. 1988. *Famine and food supply in the Graeco-Roman world: Responses to risk and crisis.* Cambridge: Cambridge University Press.
Gassón, Rafael A. 2000. Quirípas and mostacillas: The evolution of shell beads as a medium of exchange in northern South America. *Ethnohistory* 47(3–4):581–609.
Gell, Alfred. 1992. The technology of enchantment and the enchantment of technology. In *Anthropology, art and aesthetics*, edited by Jeremy Coote and Anthony Shelton, 40–66. Oxford: Clarendon.
Georgescu-Roegen, Nicholas. 1971. *The entropy law and the economic process.* Cambridge, MA: Harvard University Press.
Georgescu-Roegen, Nicholas. 1975. Energy and economic myths. *Southern Economic Journal* 41:347–381.
Georgescu-Roegen, Nicholas. 1982. The energetic theory of economic value: A topical economic fallacy. *Working Paper* No. 82-W16. Nashville: Department of Economics and Business Administration, Vanderbilt University.
Georgescu-Roegen, Nicholas. 1986. The entropy law and the economic process in retrospect. *Eastern Economic Journal* XII:3–25.
Georgescu-Roegen, Nicholas. 1991. Thermodynamics and we, the humans. Paper submitted to the *First International Conference of the European Association for Bioeconomic Studies*, Rome, November 28–30, 1991. Published in 1993 in *Entropy and bioeconomics: Proceedings of the First International Conference of the E.A.B.S.*, edited by Iosif Constantin Dragan, E. K. Seifert, and M. C. Demetrescu, 184–201. Milan: Nagard.
Glucina, Mark David, and Kozo Mayumi. 2010. Connecting thermodynamics and economics: Well-lit roads and burned bridges. *Annals of the New York Academy of Sciences* 1185:11–29.
Godelier, Maurice. 1986. *The mental and the material.* London: Verso.
Goldfrank, Walter L., David Goodman, and Andrew Szasz, eds. 1999. *Ecology and the world-system.* Santa Barbara: Greenwood Press.
Goldman, Marcio. 2009. An Afro-Brazilian theory of the creative process: An essay in anthropological symmetrization. *Social Analysis* 53(2):108–129.
Gotts, Nicholas M. 2007. Resilience, panarchy, and world-systems analysis. *Ecology and Society* 12(1):24.
Graeber, David. 2001. *Toward an anthropological theory of value: The false coin of our own dreams.* Houndmills: Palgrave.
Graeber, David. 2007. *Possibilities: Essays on hierarchy, rebellion, and desire.* Oakland: AK Press.
Graeber, David. 2011a. *Debt: The first 5,000 years.* Brooklyn: Melville House.
Graeber, David. 2011b. "Consumption." *Current Anthropology* 52(4):489–511.
Greer, John Michael. 2008. *The long descent: A user's guide to the end of the industrial age.* Gabriola Island: New Society Publishers.
Gregory, Chris. 2014. On religiosity and commercial life: Toward a critique of cultural economy and posthumanist value theory. *HAU: Journal of Ethnographic Theory* 4(3):45–68.

Grundmann, Reiner. 1991. The ecological challenge to Marxism. *New Left Review* 187:103–120.
Gudeman, Stephen. 1986. *Economics as culture: Models and metaphors of livelihood.* London: Routledge & Kegan Paul.
Gudeman, Stephen, and Alberto Rivera. 1990. *Conversations in Colombia: The domestic economy in life and text.* Cambridge: Cambridge University Press.
Gunderson, Lance H., and Crawford S. Holling, eds. 2002. *Panarchy: Understanding transformations in human and natural systems.* Washington, DC: Island Press.
Guzmán-Gallegos, María A. 2009. Identity cards, abducted footprints, and the book of San Gonzalo: The power of textual objects in Runa worldview. In *The occult life of things: Native Amazonian theories of materiality and personhood,* edited by Fernando Santos-Granero, 214–234. Tucson: The University of Arizona Press.
Håkansson, N. Thomas, and Mats Widgren, eds. 2014. *Landesque capital: The historical ecology of enduring landscape modifications.* Walnut Creek: Left Coast Press.
Hall, Charles A. S., and Kent A. Klitgaard. 2011. *Energy and the wealth of nations: Understanding the biophysical economy.* New York: Springer.
Hamilton, Clive. 2003. *Growth fetish.* London: Pluto Press.
Hanley, Nick. 1998. Resilience in social and economic systems: A concept that fails the cost-benefit test? *Environment and Development Economics* 3:244–249.
Haraway, Donna. 1991. *Simians, cyborgs, and women: The reinvention of nature.* New York: Routledge.
Harris, Marvin. 1971. *Culture, man, and nature: An introduction to general anthropology.* New York: Thomas Y. Crowell.
Hart, Keith. 2000. *Money in an unequal world.* New York: Texere.
Harvey, David. 2010. *The enigma of capital and the crises of capitalism.* Oxford: Oxford University Press.
Headrick, Daniel R. 1981. *The tools of empire: Technology and European imperialism in the nineteenth century.* Oxford: Oxford University Press.
Headrick, Daniel R. 1988. *The tentacles of progress: Technology transfer in the age of imperialism, 1850–1940.* Oxford: Oxford University Press.
Headrick, Daniel R. 2010. *Power over peoples: Technology, environments, and Western imperialism, 1400 to the present.* Princeton: Princeton University Press.
Heilbroner, Robert. 1999 [1953]. *The worldly philosophers: The lives, times and ideas of the great economic thinkers.* London: Penguin.
Heinberg, Richard. 2011. *The end of growth: Adapting to our new economic reality.* New York: New Society Publishers.
Heywood, Paolo. 2012. Anthropology and what there is: Reflections on "ontology." *Cambridge Anthropology* 30(1):143–151.
Hirschman, Albert O. 1977. *The passions and the interests: Political arguments for capitalism before its triumph.* Princeton: Princeton University Press.
Hoffmeyer, Jesper. 1996. *Signs of meaning in the universe.* Bloomington: Indiana University Press.
Horden, Peregrine, and Nicholas Purcell. 2000. *The corrupting sea: A study of Mediterranean history.* Oxford: Wiley-Blackwell.

Hornborg, Alf. 1992. Machine fetishism, value, and the image of unlimited good: Toward a thermodynamics of imperialism. *Man* (N.S.) 27:1–18.
Hornborg, Alf. 1996. Ecology as semiotics: Outlines of a contextualist paradigm for human ecology. In *Nature and society: Anthropological perspectives*, edited by Philippe Descola and Gísli Pálsson, 45–62. London: Routledge.
Hornborg, Alf. 1998. Towards an ecological theory of unequal exchange: Articulating world system theory and ecological economics. *Ecological Economics* 25(1): 127–136.
Hornborg, Alf. 2001a. *The power of the machine: Global inequalities of economy, technology, and environment*. Walnut Creek: AltaMira.
Hornborg, Alf. 2001b. Vital signs: An ecosemiotic perspective on the human ecology of Amazonia. *Sign Systems Studies* 29(1):121–152.
Hornborg, Alf. 2005. Ethnogenesis, regional integration, and ecology in prehistoric Amazonia: Toward a system perspective. *Current Anthropology* 46(4):589–620.
Hornborg, Alf. 2006. Footprints in the cotton fields: The Industrial Revolution as time-space appropriation and environmental load displacement. *Ecological Economics* 59(1):74–81.
Hornborg, Alf. 2009. Zero-sum world: Challenges in conceptualizing environmental load displacement and ecologically unequal exchange in the world-system. *International Journal of Comparative Sociology* 50(3–4):237–262.
Hornborg, Alf. 2013. *Global ecology and unequal exchange: Fetishism in a zero-sum world*. Revised paperback version. London: Routledge.
Hornborg, Alf. 2014a. Technology as fetish: Marx, Latour, and the cultural foundations of capitalism. *Theory, Culture & Society* 31(4):119–140.
Hornborg, Alf. 2014b. Political economy, ethnogenesis, and language dispersals in the prehispanic Andes: A world-system perspective. *American Anthropologist* 116(4): 810–823.
Hornborg, Alf. 2015. Why economics needs to be distinguished from physics, and why economists need to talk to physicists: A response to Foster and Holleman. *Journal of Peasant Studies* 42(1):187–192.
Hornborg, Alf, and Carole L. Crumley, eds. 2007. *The world system and the Earth system: Global socioenvironmental change and sustainability since the Neolithic*. Walnut Creek: Left Coast Press.
Hornborg, Alf, John R. McNeill, and Joan Martinez-Alier, eds. 2007. *Rethinking environmental history: World-system history and global environmental change*. Lanham: AltaMira.
Hugh-Jones, Stephen. 2009. The fabricated body: Objects and ancestors in northwest Amazonia. In *The occult life of things: Native Amazonian theories of materiality and personhood*, edited by Fernando Santos-Granero, 33–59. Tucson: The University of Arizona Press.
Ingold, Tim. 2000. *The perception of the environment: Essays in livelihood, dwelling, and skill*. London: Routledge.
Inikori, Joseph E. 1989. Slavery and the revolution in cotton textile production in England. *Social Science History* 13(4):343–379.

Inikori, Joseph E. 2002. *Africans and the Industrial Revolution in England: A study of international trade and economic development*. Cambridge: Cambridge University Press.
Jackson, Tim. 2009. *Prosperity without growth: Economics for a finite planet*. London: Earthscan.
Jacobson, Mark Z., and Mark A. Delucchi. 2011. Providing all global energy with wind, water, and solar power, Part I: Technologies, energy resources, quantities and areas of infrastructure, and materials. *Energy Policy* 39:1154–1169.
Kåberger, Tomas, and Bengt Månsson. 2001. Entropy and economic processes: Physics perspectives. *Ecological Economics* 36:165–179.
Kalb, Don. 2013. Financialization and the capitalist moment: Marx versus Weber in the anthropology of global systems. *American Ethnologist* 40(2):258–266.
Kallis, Giorgos, Joan Martinez-Alier, and Richard B. Norgaard. 2009. Paper assets, real debts: An ecological-economic exploration of the global economic crisis. *Critical Perspectives on International Business* 5(1/2):14–25.
Keen, Steve. 1993. Use-value, exchange value, and the demise of Marx's labor theory of value. *Journal of the History of Economic Thought* 15(1):107–121.
Keen, Steve. 2011. *Debunking economics: The naked emperor dethroned?* London: Zed Books.
Keil, Roger, David Bell, Peter Penz, and Leesa Fawcett, eds. 1998. *Political ecology: Global and local*. London: Routledge.
Kirchhoff, Thomas, Fridolin S. Brand, Deborah Hoheisel, and Volker Grimm. 2010. The one-sidedness and cultural bias of the resilience approach. *Gaia: Ecological Perspectives for Science and Society* 19(1):25–32.
Klein, Naomi. 2014. *This changes everything: Capitalism versus the climate*. London: Allen Lane.
Kohn, Eduardo. 2013. *How forests think: Toward an anthropology beyond the human*. Berkeley: University of California Press.
Kohn, Eduardo. 2014. Further thoughts on sylvan thinking. *HAU: Journal of Ethnographic Theory* 4(2):275–288.
Körner, Peter, Gero Maass, Thomas Siebold, and Rainer Tetzlaff. 1986. *The IMF and the debt crisis: A guide to the Third World's dilemmas*. London: Zed Books.
Latour, Bruno. 1993. *We have never been modern*. Cambridge, MA: Harvard University Press.
Latour, Bruno. 2004. *Politics of nature: How to bring the sciences into democracy*. Cambridge, MA: Harvard University Press.
Latour, Bruno. 2005. *Reassembling the social: An introduction to Actor-Network-Theory*. Oxford: Oxford University Press.
Latour, Bruno. 2009. Perspectivism: "type" or "bomb"? *Anthropology Today* 25(2):1–2.
Latour, Bruno. 2010. *On the modern cult of the factish gods*. Durham: Duke University Press.
Latour, Bruno. 2013. *Facing Gaia: Six lectures on the political theology of nature*. The Gifford Lectures on Natural Religion, Edinburgh.

Latour, Bruno. 2014. On selves, forms, and forces. *HAU: Journal of Ethnographic Theory* 4(2):261–266.
Lélé, Sharachchandra. 1998. Resilience, sustainability, and environmentalism. *Environment and Development Economics* 3:249–254.
Lenzen, Manfred, Keiichiro Kanemoto, Daniel Moran, and Arne Geschke. 2012. Mapping the structure of the world economy. *Environmental Science & Technology* 46(15):8374–8381.
Lenzen, Manfred, Daniel Moran, Keiichiro Kanemoto, and Arne Geschke. 2013. Building EORA: A global multi-regional input-output database at high country and sector resolution. *Economic Systems Research* 25(1):20–49.
Levin, Simon A., Scott Barrett, Sara Aniyar, William Baumol, Christopher Bliss, Bert Bolin, Partha Dasgupta, Paul Ehrlich, Carl Folke, Ing-Marie Gren, Crawford S. Holling, Annmari Jansson, Bengt-Owe Jansson, Karl-Göran Mäler, Dan Martin, Charles Perrings, and Eytan Sheshinski. 1998. Resilience in natural and socioeconomic systems. *Environment and Development Economics* 3:222–235.
Lévi-Strauss, Claude. 1966. *The savage mind*. Chicago: The University of Chicago Press.
Lietaer, Bernard. 2001. *The future of money: A new way to create wealth, work, and a wiser world*. London: Century.
Lipson, Daniel N. 2011. Is the Great Recession only the beginning? Economic contraction in an age of fossil fuel depletion and ecological limits to growth. *New Political Science* 33(4):555–575.
Locher, Fabien, and Jean-Baptiste Fressoz. 2012. Modernity's frail climate: A climate history of environmental reflexivity. *Critical Inquiry* 38(3):579–598.
Lonergan, Stephen C. 1988. Theory and measurement of unequal exchange: A comparison between a Marxist approach and an energy theory of value. *Ecological Modeling* 41:127–145.
Macfarlane, Alan. 1985. The root of all evil. In *The anthropology of evil*, edited by David Parkin, 57–76. Oxford: Blackwell.
MacKay, David J. C. 2012. *Sustainable energy: Without the hot air*. Retrieved June 7, 2012 from www.withouthotair.com/synopsis10.pdf
Malinowski, Bronislaw. 1954. *Magic, science and religion and other essays*. New York: Anchor Books.
Malm, Andreas. 2016. *Fossil capital: The rise of steam power and the roots of global warming*. London: Verso.
Malm, Andreas, and Alf Hornborg. 2014. The geology of mankind? A critique of the Anthropocene narrative. *The Anthropocene Review* 1:62–69.
Marcus, George E., and Michael M. J. Fischer. 1986. *Anthropology as cultural critique: An experimental moment in the human sciences*. Chicago: The University of Chicago Press.
Marglin, Stephen. 1990. Towards the decolonization of the mind. In *Dominating knowledge: Development, culture, and resistance*, edited by Frederique Apffel Marglin and Stephen Marglin, 1–28. Oxford: Clarendon.
Marsden, Ben, and Crosbie Smith. 2005. *Engineering empires: A cultural history of technology in nineteenth-century Britain*. Houndmills: Palgrave Macmillan.

Martin, Keir. 2014. Afterword: Knot-work not networks, or anti-anti-antifetishism and the ANTipolitics machine. *HAU: Journal of Ethnographic Theory* 4(3):99–115.
Martinez-Alier, Joan. 1987. *Ecological economics: Energy, environment and society*. Oxford: Blackwell.
Martinez-Alier, Joan. 1997. Some issues in agrarian and ecological economics. *Ecological Economics* 22:225–238.
Martinez-Alier, Joan. 2011. The EROI of agriculture and its use by the Via Campesina. *The Journal of Peasant Studies* 38(1):145–160.
Martinez-Alier, Joan, and Jose M. Naredo. 1982. A Marxist precursor of energy economics: Podolinsky. *Journal of Peasant Studies* 9(2):207–224.
Marx, Karl. 1967 [1867]. *Capital*, vol. I. New York: International Publishers.
Mauss, Marcel. 1990 [1925]. *The gift: The form and reason for exchange in archaic societies*. London: Routledge.
Mayumi, Kozo. 1991. Temporary emancipation from land: From the Industrial Revolution to the present time. *Ecological Economics* 4:35–56.
Mayumi, Kozo. 2009. Nicholas Georgescu-Roegen: His bioeconomics approach to development and change. *Development and Change* 40(6):1235–1254.
McAnany, Patricia A., and Norman Yoffee, eds. 2010. *Questioning collapse: Human resilience, ecological vulnerability, and the aftermath of empire*. Cambridge: Cambridge University Press.
McNally, David. 2014. The blood of the commonwealth: War, the state, and the making of world money. *Historical Materialism* 22(2):3–32.
McNeill, John R. 2000. *Something new under the sun: An environmental history of the twentieth-century world*. New York: W. W. Norton.
M'Gonigle, R. Michael. 1999. Ecological economics and political ecology: Towards a necessary synthesis. *Ecological Economics* 28:11–26.
Miller, Joana. 2009. Things as persons: Body ornaments and alterity among the Mamaindê (Nambikwara). In *The occult life of things: Native Amazonian theories of materiality and personhood*, edited by Fernando Santos-Granero, 60–80. Tucson: The University of Arizona Press.
Mintz, Sidney W. 1985. *Sweetness and power: The place of sugar in modern history*. London: Penguin.
Mirowski, Philip. 1988. Energy and energetics in economic theory: A review essay. *Journal of Economic Issues* 22(3):811–830.
Mirowski, Philip. 1989. *More heat than light: Economics as social physics, physics as nature's economics*. New York: Cambridge University Press.
Mirowski, Philip. 2013. *Never let a serious crisis go to waste: How neoliberalism survived the financial meltdown*. London: Verso.
Mitchell, Timothy. 2009. Carbon democracy. *Economy and Society* 38(3):399–432.
Mitchell, Timothy. 2011. *Carbon democracy: Political power in the age of oil*. London: Verso.
Moore, Jason W. 2015. *Capitalism in the web of life: Ecology and the accumulation of capital*. London: Verso.
Morris, Ian. 2010. *Why the West rules for now: The patterns of history and what they reveal about the future*. London: Farrar, Straus & Giroux.

Mouhot, Jean-François. 2011. Past connections and present similarities in slave ownership and fossil fuel usage. *Climatic Change* 105:329–355.
Mumford, Lewis. 1967 [1934]. *The myth of the machine: Technics and civilization.* New York: Harcourt, Brace & World.
Myrdal, Janken. 2012. Empire: The comparative study of imperialism. In *Ecology and power: Struggles over land and material resources in the past, present and future,* edited by Alf Hornborg, Brett Clark, and Kenneth Hermele, 37–51. London: Routledge.
Nadasdy, Paul. 2007. Adaptive co-management and the gospel of resilience. In *Adaptive co-management: Collaboration, learning, and multi-level governance,* edited by Fikret Berkes, Derek Armitage, and Nancy Doubleday, 208–226. Seattle: University of Washington Press.
Nader, Laura. 1996. The three-cornered constellation: Magic, science, and religion revisited. In *Naked science: Anthropological inquiry into boundaries, power, and knowledge,* 259–275. London: Routledge.
Nader, Laura. 2004. The harder path – shifting gears. *Anthropological Quarterly* 77:783–803.
Narain, Sunita, and Anil Agarwal. 1991. *Global warming in an unequal world: A case of environmental colonialism.* Delhi: Centre for Science and Environment.
Nikiforuk, Andrew. 2012. *The energy of slaves: Oil and the new servitude.* Vancouver: Greystone Books.
Norgaard, Richard B. 2013. The Econocene and the California delta. *San Francisco Estuary & Watershed Science* 11:1–5.
North, Peter. 2007. *Money and liberation: The micropolitics of alternative currency movements.* Minneapolis: University of Minnesota Press.
Nöth, Winfried. 1998. Ecosemiotics. *Sign Systems Studies* 26:332–343.
Odum, Howard T. 1988. Self-organization, transformity, and information. *Science* 242:1132–1139.
Odum, Howard T. 1996. *Environmental accounting: Emergy and environmental decision making.* New York: John Wiley & Sons.
Park, Andrew D. 2011. Beware paradigm creep and buzzword mutation. *The Forestry Chronicle* 87(3):337–344.
Parker, John N., and Edward J. Hackett. 2012. Hot spots and hot moments in scientific collaborations and social movements. *American Sociological Review* 77(1): 21–44.
Parry, Jonathan, and Maurice Bloch, eds. 1989. *Money and the morality of exchange.* Cambridge: Cambridge University Press.
Paulsen, Alison C. 1974. The Thorny Oyster and the voice of God: *Spondylus* and *Strombus* in Andean prehistory. *American Antiquity* 39:597–607.
Paulson, Susan, and Lisa Gezon, eds. 2005. *Political ecology across spaces, scales, and social groups.* Chapel Hill: Rutgers University Press.
Peet, Richard, Paul Robbins, and Michael Watts, eds. 2011. *Global political ecology.* London: Routledge.
Peet, Richard, and Michael Watts, eds. 1996. *Liberation ecologies: Environment, development, social movements.* London: Routledge.

Peirce, Charles S. 1931–1958. *Collected papers*. Cambridge, MA: Harvard University Press.
Peterson, Gary. 2000. Political ecology and ecological resilience: An integration of human and ecological dynamics. *Ecological Economics* 35:323–336.
Pfaffenberger, Bryan. 1992. Social anthropology of technology. *Annual Review of Anthropology* 21:491–516.
Pillsbury, Joanne. 1996. The Thorny Oyster and the origins of empire: Implications of recently uncovered *Spondylus* imagery from Chan Chan, Peru. *Latin American Antiquity* 7(4):313–340.
Podolinsky, Sergei. 2008 [1883]. Human labor and unity of force. *Historical Materialism* 16:163–183.
Polanyi, Karl. 1957 [1944]. *The great transformation: The political and economic origins of our time*. Boston: Beacon.
Pomeranz, Kenneth. 2000. *The great divergence: China, Europe, and the making of the modern world economy*. Princeton: Princeton University Press.
Prieto, Pedro A., and Charles A. S. Hall. 2013. *Spain's photovoltaic revolution: The energy return on investment*. New York: Springer.
Quilter, Jeffrey. 1990. The Moche revolt of the objects. *Latin American Antiquity* 1(1):42–65.
Rabinbach, Anson. 1990. *The human motor: Energy, fatigue and the origins of modernity*. New York: Basic Books.
Rappaport, Roy A. 1979. *Ecology, meaning, and religion*, 145–172. Berkeley: North Atlantic Books.
Rappaport, Roy A. 1994. Humanity's evolution and anthropology's future. In *Assessing cultural anthropology*, edited by Robert Borofsky, 153–166. New York: McGraw-Hill.
Ravenhill, John, ed. 2005. *Global political economy*. Oxford: Oxford University Press.
Reddy, William M. 1987. *Money and liberty in modern Europe: A critique of historical understanding*. Cambridge: Cambridge University Press.
Redman, Charles L. 1999. *Human impact on ancient environments*. Tucson: University of Arizona Press.
Reid, Julian. 2012. The disastrous and politically debased subject of resilience. *Development Dialogue*, April:67–79.
Roberts, J. Timmons, and Bradley C. Parks. 2007. *A climate of injustice: Global inequality, North-South politics, and climate policy*. Cambridge, MA: MIT Press.
Rotman, Brian. 1987. *Signifying nothing: The semiotics of zero*. Stanford: Stanford University Press.
Roudman, Sam. 2013. Bank of America's toxic tower: New York's "greenest" skyscraper is actually its biggest energy hog. *The New Republic*, July 28, 2013.
Sahlins, Marshall D. 1976. *Culture and practical reason*. Chicago: The University of Chicago Press.
Sahlins, Marshall D. 2013. Foreword. In *Beyond nature and culture*, by Philippe Descola, xi–xiv. Chicago: The University of Chicago Press.
Sahlins, Marshall D. 2014. On the ontological scheme of *Beyond Nature and Culture*. *HAU: Journal of Ethnographic Theory* 4(1):281–290.

Salomon, Frank. 1986. *Native lords of Quito in the age of the Incas: The political economy of North Andean chiefdoms.* Cambridge: Cambridge University Press.
Salomon, Frank, and George L. Urioste. 1991. *The Huarochirí manuscript: A testament of ancient and colonial Andean religion.* Austin: University of Texas Press.
Santos-Granero, Fernando, ed. 2009a. *The occult life of things: Native Amazonian theories of materiality and personhood.* Tucson: The University of Arizona Press.
Santos-Granero, Fernando. 2009b. Introduction: Amerindian constructional views of the world. In *The occult life of things: Native Amazonian theories of materiality and personhood*, edited by Fernando Santos-Granero, 1–29. Tucson: The University of Arizona Press.
Santos-Granero, Fernando. 2009c. *Vital enemies: Slavery, predation, and the Amerindian political economy of life.* Austin: University of Texas Press.
Santos-Granero, Fernando. 2013. Ensoulment as social agency: Life and affect at the interface between people and objects. Paper presented at the *112th Annual Meeting of the American Anthropological Association*, Chicago.
Santos-Granero, Fernando, and Frederica Barclay. 2011. Bundles, stampers, and flying gringos: Native perceptions of capitalist violence in Peruvian Amazonia. *The Journal of Latin American and Caribbean Anthropology* 16(1):143–167.
Saussure, Ferdinand de. 1916. *Cours de linguistique générale.* Lausanne: Payot.
Schaan, Denise P. 2012. *Sacred geographies of ancient Amazonia: Historical ecology of social complexity.* Walnut Creek: Left Coast Press.
Scheidel, Walter, ed. 2009. *Rome and China: Comparative perspectives on ancient world empires.* Oxford: Oxford University Press.
Schrödinger, Erwin. 1967 [1944]. *What is life? Mind and matter.* Cambridge: Cambridge University Press.
Schumacher, Ernst Friedrich. 1974. *Small is beautiful: A study of economics as if people mattered.* London: Abacus.
Schwartz, Stuart B. 2011. The Iberian Atlantic to 1650. In *The Oxford handbook of the Atlantic world 1450–1850*, edited by Nicholas Canny and Philip Morgan, 147–164. Oxford: Oxford University Press.
Schwartzman, David. 1996. Solar communism. *Science & Society* 60(3):307–331.
Schwartzman, David. 2008. The limits to entropy: Continuing misuse of thermodynamics in environmental and Marxist theory. *Science & Society* 72(1):43–62.
Screpanti, Ernesto. 2014. *Global imperialism and the great crisis: The uncertain future of capitalism.* New York: Monthly Review Press.
Sebeok, Thomas A., and Jean Umiker-Sebeok, eds. 1992. *Biosemiotics.* Berlin: Mouton de Gruyter.
Sen, Amartya. 1987. *On ethics and economics.* Oxford: Blackwell.
Sheridan, Michael J. 2012. Water: Irrigation and resilience in the Tanzanian highlands. In *Ecology and power: Struggles over land and material resources in the past, present, and future*, edited by Alf Hornborg, Brett Clark, and Kenneth Hermele, 168–181. London: Routledge.
Sieferle, Rolf P. 2001 [1982]. *The subterranean forest: Energy systems and the Industrial Revolution.* Cambridge: White Horse Press.

Simas, Moana, Richard Wood, and Edgar Hertwich. 2015. Labor embodied in trade: The role of labor and energy productivity and implications for greenhouse gas emissions. *Journal of Industrial Ecology* 19(3):343–356.
Simmel, Georg. 1990 [1907]. *The philosophy of money*. London: Routledge.
Smil, Vaclav. 1992. Elusive links: Energy, value, economic growth and quality of life. *OPEC Review* 16:1–21.
Smil, Vaclav. 1994. *Energy in world history*. Boulder: Westview Press.
Smil, Vaclav. 2006. *Energy: A beginner's guide*. Oxford: Oneworld.
Smith, Michael E. 2001. The Aztec Empire and the Mesoamerican world system. In *Empires: Perspectives from archaeology and history*, edited by Susan E. Alcock, Terence N. D'Altroy, Kathleen D. Morrison, and Carla M. Sinopoli, 128–154. Cambridge: Cambridge University Press.
Smith-Nonini, Sandy. 2014. Oil, debt, and neoliberal rule in the United States. Paper presented at meeting of the *Society for Economic Anthropology*, Austin, April 24–26, 2014.
Soddy, Frederick. 1926. *Wealth, virtual wealth and debt*. London: Allen & Unwin.
Söderberg, Johan, and Adam Netzén. 2010. When all that is theory melts into (hot) air: Contrasts and parallels between actor network theory, autonomist Marxism, and open Marxism. *Ephemera* 10(2):95–118.
Solar energy. n.d. In *Wikipedia*. Retrieved September 2, 2011 from http://en.wikipedia.org/wiki/Solar_energy
Solar power. n.d. In *Wikipedia*. Retrieved September 2, 2011 from http://en.wikipedia.org/wiki/Solar_power
Stevis, Dimitris, and Valerie J. Assetto, eds. 2001. *The international political economy of the environment: Critical perspectives*. Boulder: Lynne Rienner.
Stiglitz, Joseph E. 2010. *Freefall: America, free markets, and the sinking of the world economy*. New York: W. W. Norton.
Strange, Susan. 1994. From Bretton Woods to the casino economy. In *Money, power and space*, edited by Stuart Corbridge, Ron Martin, and Nigel Thrift, 49–62. Oxford: Blackwell.
Strum, Shirley, and Bruno Latour. 1987. Redefining the social link: From baboons to humans. *Social Science Information* 26:783–802.
Tainter, Joseph A. 1988. *The collapse of complex societies*. Cambridge: Cambridge University Press.
Tainter, Joseph A., and Tadeusz W. Patzek. 2012. *Drilling down: The Gulf oil debacle and our energy dilemma*. New York: Springer.
Tambiah, Stanley J. 1990. *Magic, science, religion, and the scope of rationality*. Cambridge: Cambridge University Press.
Taussig, Michael T. 1980. *The devil and commodity fetishism in South America*. Chapel Hill: University of North Carolina Press.
Torres, Constantino M. 1987. *The iconography of South American snuff trays*. Etnologiska Studier 37. Gothenburg: Göteborgs Etnografiska Museum.
Townsend, Erik. 2013. Why peak oil threatens the international monetary system. *Association for the Study of Peak Oil and Gas (ASPO)*. http://www.resilience.org/stories/2013-01-07/commentary-why-peak-oil-threatens-the-international-monetary-system [accessed November 29, 2014].

Trawick, Paul, and Alf Hornborg. 2015. Revisiting the image of limited good: On sustainability, thermodynamics, and the illusion of creating wealth. *Current Anthropology* 56(1):1–27.

Turner, Terence. 2009. Valuables, value, and commodities among the Kayapó of central Brazil. In *The occult life of things: Native Amazonian theories of materiality and personhood*, edited by Fernando Santos-Granero, 152–169. Tucson: The University of Arizona Press.

Victor, Peter A. 2008. *Managing without growth: Slower by design, not disaster*. Cheltenham: Edward Elgar.

Vigh, Henrik E., and David B. Sausdal. 2014. From essence back to existence: Anthropology beyond the ontological turn. *Anthropological Theory* 14(1):49–73.

Viveiros de Castro, Eduardo. 1998. Cosmological deixis and Amerindian perspectivism. *The Journal of the Royal Anthropological Institute* 4(3):469–488.

Viveiros de Castro, Eduardo. 1999. The transformation of objects into subjects in Amerindian ontologies. Paper presented at the *98th Annual Meeting of the American Anthropological Association*, Chicago.

Von Bertalanffy, Ludwig. 1968. *General system theory: Foundations, development, applications*. Harmondsworth: Penguin Books.

Von Thünen, Johann Heinrich. 1966 [1826]. *Von Thünen's Isolated State*. New York: Pergamon.

Von Uexküll, Jakob. 1982 [1940]. The theory of meaning. *Semiotica* 42:25–82.

Wackernagel, Mathis, and William E. Rees. 1996. *Our ecological footprint: Reducing human impact on the Earth*. Gabriola Island: New Society Publishers.

Walker, Harry. 2009. Baby hammocks and stone bowls: Urarina technologies of companionship and subjection. In *The occult life of things: Native Amazonian theories of materiality and personhood*, edited by Fernando Santos-Granero, 81–102. Tucson: The University of Arizona Press.

Walker, Jeremy, and Melinda Cooper. 2011. Genealogies of resilience: From systems ecology to the political economy of crisis adaptation. *Security Dialogue* (Special issue on *The Global Governance of Security and Finance*) 42(2):143–160.

Wallerstein, Immanuel. 1974–1989. *The modern world-system*, 1–3. San Diego: Academic Press.

Warlenius, Rikard. 2011. *Iron for tea: The trade of the Swedish East India Company as a cross-continental case study of ecologically unequal exchange in the eighteenth century*. Master Thesis, Department of Economic History, Stockholm University.

Weatherford, Jack. 1997. *The history of money: From sandstone to cyberspace*. New York: Three Rivers Press.

White, Leslie A. 1959. *The evolution of culture: The development of civilization to the fall of Rome*. New York: McGraw-Hill.

Widgren, Mats. 2007. Precolonial landesque capital: A global perspective. In *Rethinking environmental history: World-system history and global environmental change*, edited by Alf Hornborg, John R. McNeill, and Joan Martinez-Alier, 61–77. Lanham: AltaMira Press.

Widgren, Mats. 2012. Resilience thinking versus political ecology: Understanding the dynamics of small-scale, labour-intensive farming landscapes. In *Resilience and the cultural landscape: Understanding and managing change in human-shaped environments*, edited by Tobias Plieninger and Claudia Bieling, 95–110. Cambridge: Cambridge University Press.

Wilding, Adrian. 2010. Naturphilosophie redivivus: On Bruno Latour's "political ecology." *Cosmos and History: The Journal of Natural and Social Philosophy* 6:1.

Wilk, Richard R., and Lisa C. Cliggett. 2007. *Economies and cultures: Foundations of economic anthropology*. Boulder: Westview Press.

Wilkinson, Richard G. 1973. *Poverty and progress: An ecological model of economic development*. London: Methuen.

Williams, Michael. 2003. *Deforesting the Earth: From prehistory to global crisis*. Chicago: The University of Chicago Press.

Williams, Michael. 2007. The role of deforestation in Earth and world-system integration. In *Rethinking environmental history: World-system history and global environmental change*, edited by Alf Hornborg, John R. McNeill, and Joan Martinez-Alier, 101–122. Lanham: AltaMira Press.

Woolf, Greg. 2012. *Rome: An empire's story*. Oxford: Oxford University Press.

World Commission on Environment and Development. 1987. *Our common future*. Oxford: Oxford University Press.

Worsley, Peter. 1970 [1957]. *The trumpet shall sound: A study of "cargo" cults in Melanesia*. London: Paladin.

Wrigley, Edward Anthony. 1962. The supply of raw materials in the Industrial Revolution. *Economic History Review Series* 2(15):1–16.

Yergin, Daniel. 1991. *The prize: The epic quest for oil, money, and power*. New York: Free Press.

Yoffee, Norman, and George L. Cowgill, eds. 1988. *The collapse of ancient states and civilizations*. Tucson: University of Arizona Press.

York, Richard. 2012. Do alternative energy sources displace fossil fuels? *Nature Climate Change* 2:441–443.

Yu, Yang, Kuishuang Feng, and Klaus Hubacek. 2013. Tele-connecting local consumption to global land use. *Global Environmental Change* 23(5):1178–1186.

Yule, Henry. 1871. *The book of Ser Marco Polo, the Venetian*. London: John Murray.

Zehner, Ozzie. 2012. *Green illusions: The dirty secrets of clean energy and the future of environmentalism*. Lincoln: University of Nebraska Press.

Index

Actor-Network Theory (ANT), 10, 11, 15, 34, 107, 163, 172n10
Africa, 62
agency, 96, 99
agriculture, 126, 168–69n4
Albritton Jonsson, Fredrik, 42
Algeria, 125
Amazonian societies, 101–2
   human-object relations in, 105–8
Amin, Samir, 83
analogism, 97–98, 101, 102
Andean societies, 101–2, 148
animism, 97, 101, 102, 111
Anthropocene (epoch), 16, 18, 33–34, 163
   alternatives to, 166n7
   sustainability in, 27–29
anthropology, 2–3, 5, 95
   economic, 93
   economics and, 46
   ontology in, 96–99
   relational approaches in, 11–14
   resilience theory and, 136–37
   of sustainability in Anthropocene, 27–29
   symmetric anthropology, 105
   on technology, 18
   on value, 26, 166n1
Aquinas, Thomas, 44, 45, 154
Argentina, 129, 133
Århem, Kaj, 98
Aristotle
   on chrematistics, 154
   on money, 25, 43, 44, 46, 48, 129, 135, 161–62
   on use value, 41, 82, 174n1

artifacts
   Actor-Network Theory on, 10
   agency attributed to, 96, 99
   fetishism of, 94
   in human-object relations, 103–8
   symbolic significance of, 93
   technology as, 158
assemblage, 99
avarice, 44
Axial Age, 47
Ayres, Robert, 121–22
Aztec Empire, 61, 62

baboons, 10
Bank of America, 113
Barasana (people), 105
Bateson, Gregory, 11–14, 142
Baudrillard, Jean, 82
Benton, Ted, 123
Best, Elsdon, 93
biogeographical models, 55–56
Bird-David, Nurit, 109
Bloch, Maurice, 44
Bohannan, Paul, 46, 143
Brand, Fridolin Simon, 174n6
Braudel, Fernand, 17
Bretton Woods Agreement, 47, 51, 168n11
British East India Company, 62
British Empire, 62–63, 73
   imperial strategy of, 74
   trade of, 64–65
Brundtland report (World Commission on Environment and Development), 117

Bunker, Stephen, 27
Burbank, Jane, 66
Burkett, Paul, 80–83, 85–87, 124, 160
Byzantine Empire, 60

capital, 67–68
  asymmetric appropriation of, 79
  definition of, 151
  in premodern societies, 19
  unequal exchange and, 69
capitalism, 33–34, 151, 161, 166n7
  early merchant capitalism, 155
  fetishism in, 108–11
  financialization in, 48, 50–51
  fossil fuels used in, 170n12
  independent of human biology, 36
  money and, 174n2
  morality and, 44
  Polanyi on, 46
  tied to steam power, 152
carbon capture and storage, 166n9
carbon dioxide emissions, 33
cargo cults, 1–2, 5
Caribbean Sea, 62
Cartesian dualism, 13, 30–31
Chakrabarty, Dipesh, 34
China
  Han dynasty and empire in, 59
  landscapes in, 71
Chipaya (people), 101
chrematistics, 157, 174n1
  Aristotle on, 25, 43, 48, 154
  physics and, 169n6
Christianity
  in Roman Empire, 60
  in Spanish Empire, 61
Clark, Brett, 57–58
climate change, 33, 167n5
coal, 65
coins, 104, 171n5
colonialism, 32
commensurability
  resilience theory and, 143–49
  of values, 131–36
communication, 19
  semiotics as study of, 39
  social systems as, 13

complementary currencies, 132–35, 145–46, 159
Conant, James D., 117
Constantinople, 60
consumerism, 161
Cooper, Frederick, 66
core, of world-system, 14
Corn Laws (U.K.), 42, 156, 157
cosmology, 3
  see also ontology, worldview
Costanza, Robert, 168n3
cotton, 17, 21–23
credit, 131
  consumption tied to, 50
credit money, 47, 48
Crosby, Alfred, 55–58, 71
Crutzen, Paul, 32
culture, 140
currencies
  alternative, 130–31
  complementary, 132–35, 159, 173n1
  local, 145–46
  metal coins for, 46–47
  paper used for, 37–38, 47
  see also money
Cuzco, 60
cybernetics, 11–12

Dalton, George, 46
Daniels, Farrington, 117
Delucchi, Mark A., 119
Descartes, René, 30–31
Descola, Philippe, 96–98, 100, 102–4, 106–7
Diamond, Jared, 55–56, 74
digital money, *see* electronic money
disciplines (academic), 8
displacement of work and
  environmental loads, 17, 32, 35, 42, 115, 156–57, 165n2–3, 167n6
dollar, 51, 168n11
Duffield, Mark, 146–47
Dumont, Louis, 44
Dutch East India Company, 61

ecological economics, 25
  and eco-Marxists, 78

on entropy, 120
on exchange values and use
    values, 157
on financial crisis, 51
Marxism and, 123–24
on material aspects of economic
    activity, 84
on money, 41, 131
on technology, 126
on unequal exchange, 82, 88
ecological imperialism, 57–58, 68–70,
    73–74
ecologically unequal exchange, as
    asymmetric flows of biophysical
    resources, 70, 84, 152
ecology, 7, 13
anthropology and, 97
and imperialism, 54–58
political ecology, 34–36, 165n5
of social–ecological systems, 139–41
eco-Marxists, 78, 160
economic anthropology, 93
economics
anthropological view of, 5
emergence of, 40–43
establishment of discipline of, 45–46
morality and, 44, 154
neoclassical, 43
as social utilization of energy
    sources, 10
thermodynamics and, 120–23
unequal exchange in, 153
ecosemiotics, 107
efficiency, 116
Egypt, ancient, 9, 13
Elder-Vass, Dave, 170–71n2
electronic money, 47, 135, 174n4
Elizabeth (queen, England), 49
Emmanuel, Arghiri, 22, 25, 83
empires
comparisons of, 58–63
continuities and discontinuities in,
    66–68
flow of money in, 88
roles of money and technology in,
    70–73
see also imperialism

energy, 155
economic theories on, 84–87
entropy of, 120, 122
flowing from sun, 18–19
flows of, 88–91
fossil fuels for, 17
in living things, 15
in metabolism of empires, 73–74
money and, 20–21, 76–80, 160
return on energy investment
    (EROI), 77
social utilization of, 9–10
solar power for, 117–20, 124–27
in theory of value, 75, 76, 82
energy imperialism, 58
Engels, Friedrich, 75–76, 86, 160, 168n3
engineering, 9, 13, 17, 32, 125
Enlightenment, 103, 104
entropy, 15, 26, 121–24, 165n1
see also Second Law of
    Thermodynamics
environment, 7
solar power as benign for, 117–20
environmental load displacement, 157,
    165n3
see also displacement
Erikson, Philippe, 105
Europe
financial crises in, 147
imperialistic expansion of, 56–57
labor energy in, 141
exchange values, 41–42, 79
Aristotle on, 174n1
paper money for, 38
use values versus, 82, 157
exploitation, 4, 31, 43, 50, 60, 68,
    83–84

Fausto, Carlos, 172n9
fetishism, 14
as attribution of power, 172n13
in capitalist modernity, 108–11
in human-object relations, 104
Marx on, 6, 11, 95, 105, 107–8
money as, 21, 72, 94, 135
posthumanist, 171n8
field theories of technology, 14–16

financial crises, 46–51, 129–30, 147
  complementary currencies in, 133
Financial Crisis Inquiry Commission, 49
financial imperialism, 71, 147
financialization, 48, 50–51, 131, 168n15
First Law of Thermodynamics, 76
Fischer, Michael M.J., 95, 172n10
Fortun, Kim, 172n10
fossil fuels, 15, 16, 22, 116–17
  capitalism's shift to, 156, 170n12
  in metabolism of empires, 73–74
  solar power as alternative to, 118–20
  used by steam engines, 17, 18
Foster, George, 3
Foster, John Bellamy, 168–69n4
  on energy imperialism, 57–58
  on financialization, 50
  on labor-power and energy, 83, 85
  on Podolinsky, 86–87, 160
Foucault, Michel, 8
Frank, Andre Gunder, 14, 56–57

Gell, Alfred, 172n14
Georgescu-Roegen, Nicholas, 90, 166n2, 167n5
  criticisms of, 80–81, 124
  on economics and thermodynamics, 41, 77, 79, 120–23, 125, 126, 159, 170n10
  on money, 130
general-purpose money, 18, 33, 35–6, 39, 43–6, 79–80, 90, 136, 142
globalization, 142–43, 173n16
  complementary currencies as opposite of, 135
Godelier, Maurice, 68
gold, 37–8, 47–8, 53, 61
Goldman, Marcio, 108
gold standard, 47–48, 130
Gore, Al, 113
Gotts, Nicholas M., 137, 142–43
Graeber, David, 21, 44, 106, 167n8
Greece
  ancient, 46–47, 82
  current, 129, 133
Gregory, Chris, 171n8, 172n15

growth, economic, 16, 18, 22, 55, 64, 77, 88, 90
Grundmann, Reiner, 123–24
Gunderson, Lance H., 146
Guzmán-Gallegos, María, 108

Habsburg dynasty, 61
Hall, Charles A.S., 118
Han dynasty (China), 59
Harris, Marvin, 2
Harvey, David, 40
Hatshepsut (queen, Egypt), 9, 13, 56
Heilbroner, Robert, 45, 46, 131, 166n2, 167n3
Hobson, John, 43
Holling, Crawford, 137, 139, 141, 142, 146, 148
Howell, Signe, 97
Hudson's Bay Company, 63
Hugh-Jones, Stephen, 100, 105
humanism, 163
human-object relations, 103–4
  objectification and subjectification in, 105–8
humans
  Anthropocene epoch and, 33–34
  as efficient machines, 75
  environmental impact of, 28–29, 165n5
  fossil fuels used by, 76
  Latour on, 99
  as part of biosphere, 163
  social inequalities among, 10
  social metabolism of, 18–20
  steam engines replacing labor of, 63–64
hydroelectric power, 116–17

identity, concept of, 138–39
imperialism
  comparisons of empires, 58–63
  continuities and discontinuities in, 66–68
  ecological, 55–58, 68–70, 73–74
  ecology of, 54–55
  financial, 71

globalization as replacement for, 173n16
Inca Empire, 53, 60–61
   landscapes in, 71
   power in, 140–41
   Spanish conquest of, 62
   tribute in, 70
India, 63
indigenous people, 38
industrialization, 33–34, 63–66, 157
Industrial Revolution, 42
   in British Empire, 63
   global transfers of resources during, 65
   human energy replaced during, 63–64
   money necessary for, 21, 36
   naturalism and, 102
   post-Cartesian understanding of, 35
   steam energy for, 17, 18, 155
   transformation of technology during, 31–33
inequality, 10
Ingold, Tim, 12–14
intentionality, 104, 108–09, 171n8
International Monetary Fund (IMF), 129

Jacobson, Mark Z., 119
Jamaica, 63
Jax, Kurt, 174n6
Jevons, William Stanley, 43, 114

Kåberger, Tomas, 122, 123
Kayapó (people), 105
Keynes, John Maynard, 43, 104
keys, 104
Kohn, Eduardo, 107
Kublai Khan, 37

labor, 54
labor-power, 78
   energy and, 85
   in export production, 170n12
labor theory of value, 42–43, 46, 78, 86, 87, 169–70n7
land, 19–20
   classical economists on limits to, 65
   economic theories on, 84
   as factor of production, 152
   labor invested in, 54
   Physiocrats on, 82
landscapes, 71–72
language, 140
Latour, Bruno, 162, 172n10
   Actor-Network Theory of, 10, 11, 15, 34
   on artifacts in social organization, 102
   criticisms of, 166n8, 170–71n2, 171n3
   on fetishism, 107
   on modernity, 103
   ontology of, 96, 98–99
   political ecology of, 35
   symmetric anthropology of, 105
Leadership in Energy and Environmental Design (LEED), 113
Lélé, Sharachchandra, 137
Lévi-Strauss, Claude, 100, 137
Lietaer, Bernhard, 135
local currencies, 130, 143–45, 173n2
Local Exchange Trading Systems (LETS), 132–35, 143
Lonergan, Stephen C., 82–83, 88

machines, 109
MacKay, David, 118
Magdoff, Fred, 50
magic, 5–6, 94–95
   conventionally distinguished from technology, 103
   technology as, 108–11, 159–60
Malinowski, Bronislaw, 6
Malm, Andreas, 166n7
Malthus, Thomas, 22, 64, 82, 114
Mandeville, Bernard, 44, 154
Månsson, Bengt, 122, 123
Maori (people), 10–11, 93, 94
Marcus, George, 95
market, 40–42
   principle, 30, 46, 94–95, 141
   prices prerequisite to technology, 7, 16, 21–23, 32, 65–67
   scale, 46, 135
Marshall, Alfred, 41

Martinez-Alier, Joan, 160
Marx, Karl, 5
  on artifacts, 96
  on capital, 151
  on commodities, 11
  on depletion of soils, 80–81
  on energy, 85
  on fetishism, 6, 7, 14, 95, 105, 107–8, 154, 159–60
  labor theory of value of, 42–43, 82, 83, 169n5
  on money, 21, 41, 72, 94, 109, 149
  on Physiocracy, 85–86
  Podolinsky and, 75, 86–87, 160
  on rural life, 168–69n4
  on technological progress, 22, 64
Marxism
  on capital accumulation, 77
  ecological economics and, 123–24
  ecological issues and, 80–81, 169–70n7
  eco-Marxists, 78
  on economic growth, 88
  entropy and, 120–21
  on environmental problems, 90
  on financial crisis, 50
  on flows of labor-power, 25
  labor theory of value in, 78, 87
  on material aspects of economic activity, 84
  on money, 44, 131
  on profits, 48
  technological utopianism in, 114
  on technology, 126
  thermodynamics and, 75–76, 160
  on unequal exchange, 68–69, 83, 88, 91
  on use values, 82, 157
  on value, 26
Matis (people), 105, 171n6
matter, entropy of, 120–22
Mauss, Marcel, 10–11, 18, 93, 108
Maya (people), 138–39
Mayumi, Kozo, 166n2
Melanesia, 1–2
mercantile capitalism, 88, 155
  empires tied to, 66–67, 69

metabolism
  of empires, 58–63, 72, 73
  social, 18–20, 145
Mexico, 3, 61
Middle Ages, 47
Miller, Joana, 105
Mitchell, Timothy, 40
Moche (people), 106
modernity, 38
  modern versus nonmodern, 1, 5–7, 43, 96–98, 106, 109, 173n17
  premodern versus modern, 2, 6, 10, 19, 93–94, 103
money, 95, 149
  accepted in all economic theories, 79
  alternative currencies, 130–31
  as condition for technology, 20–23
  in early modern Europe, 155
  in ecology of empires, 70–73
  as fetishism, 6, 94, 161–62, 174n2
  as fictive energy, 76–80
  financial crises and, 129–30
  flows of, 88–91
  global distribution of, 16
  Marx on, 109
  metal coins used for, 46–47
  morality and, 43–46, 154
  necessary for Industrial Revolution, 36
  paper used for, 37–38, 47
  resilience theory and, 136, 142
  semiotics of, 38–40
  value and, 40–43
  see also general-purpose money
Moore, Jason, 169–70n7
morality
  economics and, 154
  money and, 43–46
Mughal Empire, 62
Mumford, Lewis, 13

Nadasdy, Paul, 137
nations, 142
naturalism, 102–3, 111
natural resources
  agriculture dependent upon, 126
  consumption of, 81

nature
  anthropological study of, 97
  society and, 29–33
neoclassical economics, 43
neo-Physiocracy, 81, 82, 84, 90
Netherlands, 61
nuclear power, 117, 166n9

objects
  perceived as subjects, 108
  revolt of, 106
  see also fetishism; human-object relations
Odum, Howard T., 81, 82, 86, 121, 170n13
ontology, 96–103
Owen, Robert, 167n3

Pálsson, Gísli, 97
paper money, 37–38, 47
Parry, Jonathan, 44
Paul (saint), 43–44, 135, 154
Peirce, Charles Sanders, 39
periphery, of world-system, 14
personhood, 104, 107, 109, 171n3
perspectivism, 98
photovoltaics, 117–20, 124, 125
Physiocrats and Physiocracy, 41–43, 81, 82
  on economic growth, 88
  Marx on, 85–86
  on material aspects of economic activity, 84
Podolinsky, Sergei, 75–76, 86–87, 122, 160, 170n13
Polanyi, Karl, 45–46
political ecology, 165n5
  resilience theory and, 139
political economy, 96, 99–103
Polo, Marco, 37
Pomeranz, Kenneth, 56–57, 64–65
Portugal, 62
power, social, 140–41, 162
precious metals, 37, 47, 53, 62, 135
  see also gold, silver
premodern societies, 19–20
prestige-good systems, 100

Prieto, Pedro A., 118
progress, 70
proto-machines, 73

Quesnay, François, 42
Quilter, Jeffrey, 106

Rabinbach, Anson, 85, 168n2
racism, 4
Ranaipiri, Tamati, 93
Rappaport, Roy, 27, 142, 163
rationality, 5–6, 21, 23
recycling, 121–22
Reid, Julian, 146–47
relationism, 7
resilience theory, 135–39, 141–43, 174n6
  commensurability and, 143–49
Ricardo, David
  on capital and labor substituting for land, 22, 64, 65, 156–57
  labor theory of value of, 41, 42, 82, 123
Roman Empire, 59–60, 73
Roudman, Sam, 113
Royal Africa Company, 62–63
Ruiz, Bartolomé, 53

Sahlins, Marshall, 27, 82, 97, 102
Santos-Granero, Fernando, 100, 102, 104
  on subjectivity of objects, 106, 108
Saussure, Ferdinand de, 39
Scandinavian countries, 147
Schumpeter, Joseph, 46
Schwartzman, David, 80–81, 124, 147–48
science, 6
  natural versus social, 28, 34
science and technology studies (STS), 29, 98, 107
scientific revolution, 171n4
Second Law of Thermodynamics, 18, 26, 79–80, 165n1
  economics and, 120, 122
semiotics, 39
Shakespeare, William, 44
Sieferle, Rolf Peter, 65

signs, 39
silver, 53, 62
Simas, Moana, 25, 165n2
slavery and slaves, 44
   in ancient Egypt, 9–10, 13
   British trade in, 63
   cotton and, 17, 21
   Industrial Revolution tied to, 155
   Portuguese trade in, 62
   technological change tied to, 32
   technology as displacement of, 23–25
   wage labor and, 167n8
Smil, Vaclav, 117, 118, 121
Smith, Adam, 42, 44, 82, 154
social–ecological systems, 139–41
social metabolism, 18–20
social sciences, 28, 34
societies
   nature and, 29–33
   political power in, 35
   social–ecological systems and, 139–41
Soddy, Frederick, 51, 167n4
solar power, 114, 117–20, 123, 124
   technology of, 125–27
Spain, 53, 173n3
Spanish Empire, 61–62
Spondylus (Thorny Oyster), 38, 53, 100–102, 109, 110
stagnation, 50, 168n15
steam engines
   British Empire's use of, 63
   Crutzen on, 32
   development of capitalism tied to, 152
   human energy replaced by, 63–64
   invention of, 17, 18, 155
   long-term effects of, 113
   mode of production tied to, 151
   slave trade tied to, 24
Stiglitz, Joseph, 49–50, 168n12–13
Strum, Shirley, 10, 99, 162
subjectivity, 96, 98, 102–103, 106–108
surplus production, 85–87
sustainability, anthropology on, 27–29
Sweden, 64
Sweezy, Paul, 50
symbolic systems, 140
symmetric anthropology, 105

Taussig, Michael, 6, 108
technique, 12–13
Technocene, 34–36
Technocrat movement, 130, 170n8
technofetishism, 7
technological development, 17
technology, 94
   of ancient Egypt, 9
   anthropology of, 18
   as artifact, 158
   as displacement of slavery, 23–25
   conventionally distinguished from magic, 6, 103
   in ecology of empires, 70–73
   field theory of, 14–16
   flows of resources in, 151–52
   Industrial Revolution transformation of, 31–32
   as magic, 108–11, 159–60
   Marx on, 64
   Marxism and, 123–24
   money as condition for, 20–23
   in political economy, 96
   progress in, as cultural category, 114–17
   resilience theory on, 146
   of solar power, 117–20, 125–27
   as systems of artifacts, 10
   technique distinguished from, 12–13
   technofetishism and, 7
   theories on progress in, 81
technomass, 12, 31, 33
Tenochtitlan, 61
thermodynamics
   economics and, 120–23
   on labor-power, 83, 85
   laws of, 30, 76, 77
   limits to growth based on, 125
   Podolinsky on, 75–76, 160
   Second law of, 18, 26, 79–80, 165n2
Thorny Oyster, *see Spondylus*
time–space appropriation, 152
transdisciplinarity, 4, 7–8, 18, 27, 29, 31
triangular trade, 63, 155
tribute, 69–70
Túpac Amaru revolt, 62

Turgot, Anne Robert Jacques, 85
Turner, Terry, 105

Umayyad caliphate (Iberia), 61
underpayment, 78–79, 83–84, 88–90
United Kingdom, as British Empire, 62–63
United States
  British Empire replaced by, 63
  financial crises in, 147
  imperial strategy of, 74
use values, 41, 78–84, 158
  Aristotle on, 174n1
  exchange values versus, 157
utility, 157
Utopianism, technological, 35, 114, 124–25

values (economic), 166n1
  commensurability of, 131–36
  entropy and, 122

money and, 38, 40–43
  as mystification, 25–27
  unequal exchanges and, 69, 158
Veblen, Thorstein, 45
Virginia Company, 62
Viveiros de Castro, Eduardo, 98, 99, 107
von Thünen, Johann Heinrich, 145

Wachtel, Nathan, 101
Walker, Harry, 107
Wallerstein, Immanuel, 14, 15
water mills, 24
Watt, James, 17, 18, 32, 155
White, Leslie, 74
World Commission on Environment and Development, 117
world-system analysis, 14
worldview, 2, 4, 28, 32–33, 35
Worsley, Peter, 1–2

York, Richard, 118

The manufacturer's authorised representative in the EU is Springer Nature Customer Service Centre GmbH, Europaplatz 3, 69115 Heidelberg, Germany. If you have any concerns regarding our products, please contact ProductSafety@springernature.com

Printed and bound by CPI Group (UK) Ltd, Croydon, CR0 4YY
23/03/2026
02076449-0014